OUR CHANGING PLANET

THE FY 2001
U.S. GLOBAL CHANGE RESEARCH PROGRAM

A Report by the Subcommittee on Global Change Research,
Committee on Environment and Natural Resources
of the National Science and Technology Council

A Supplement to the President's Fiscal Year 2001 Budget

About the National Science and Technology Council

President Clinton established the National Science and Technology Council (NSTC) by Executive Order on November 23, 1993. This cabinet-level council is the principal means for the President to coordinate science, space, and technology policies across the Federal Government. The NSTC acts as a "virtual" agency for science and technology to coordinate the diverse parts of the Federal research and development enterprise. The NSTC is chaired by the President. Membership consists of the Vice President, the Assistant to the President for Science and Technology, Cabinet Secretaries and Agency Heads with significant science and technology responsibilities, and other senior White House officials.

An important objective of the NSTC is the establishment of clear national goals for Federal science and technology investments in areas ranging from information technology and health research, to improving transportation systems and strengthening fundamental research. The Council prepares research and development strategies that are coordinated across Federal agencies to form an investment package that is aimed at accomplishing multiple national goals.

To obtain additional information regarding the NSTC, contact the NSTC Executive Secretariat at 202-456-6100 (voice).

About the Committee on Environment and Natural Resources

The Committee on Environment and Natural Resources (CENR) is one of nine committees under the NSTC, and is charged with improving coordination among Federal agencies involved in environmental and natural resources research and development, establishing a strong link between science and policy, and developing a Federal environment and natural resources research and development strategy that responds to national and international issues.

To obtain additional information about the CENR, contact the CENR Executive Secretary at 202-482-5917 (voice).

About the Office of Science and Technology Policy

The Office of Science and Technology Policy (OSTP) was established by the National Science and Technology Policy, Organization, and Priorities Act of 1976. OSTP's responsibilities include advising the President on policy formulation and budget development on all questions in which science and technology are important elements; articulating the President's science and technology policies and programs; and fostering strong partnerships among Federal, State, and local governments, and the scientific communities in industry and academia.

To obtain additional information regarding the OSTP, contact the OSTP Administrative Office at 202-456-6004 (voice).

EXECUTIVE OFFICE OF THE PRESIDENT
OFFICE OF SCIENCE AND TECHNOLOGY POLICY
WASHINGTON D.C. 20502

September 2000

Members of Congress:

I am pleased to transmit to you a copy of *Our Changing Planet: The FY 2001 U. S. Global Change Research Program.* This report, prepared under the auspices of the President's National Science and Technology Council (NSTC), highlights the Program's recent research and describes future plans and goals. The U.S. Global Change Research Program (USGCRP) was established in 1989 and authorized by Congress in the Global Change Research Act of 1990. The first edition of *Our Changing Planet* was transmitted to the Congress as a supplement to the FY 1990 budget. In just over a decade, the USGCRP has generated remarkable improvements to our knowledge of Earth's global-scale environmental processes and helped identify and explain the causes and consequences of a series of global environmental changes, including ozone depletion and climate change.

In a natural outgrowth of the progress in global-scale analysis, the USGCRP will soon complete its first National Assessment of the Potential Consequences of Climate Variability and Change, which describes the projected impacts of climate change and variability in the United States on a regional scale. This effort is the culmination of more than two years of work by hundreds of scientists from around the country. The National Assessment will provide managers, planners, organizations, and the public with the information needed to increase our resilience to climate variability and to improve our ability to cope with climate change. The production of this document is an important milestone for the USGCRP. As we look ahead to the next decade for the program, the creation of useful scientific products that contribute to the information needs of decision-makers is expected to be a more prominent part of this important research effort.

The assessment also identifies key research needs for continuing improvements in understanding climate impacts. As such, it has been an important contribution to the development of a new long-term strategy for the USGCRP. This document, which will be submitted to the Congress as required in the authorizing legislation of the USGCRP, will provide a long-term vision for the continuing development of our nation's global change research efforts. A strong continuing base of physical science research will be augmented by a greater focus on biological, ecological, and social science research focused on the accurate characterization of the vulnerability and resilience of natural and managed ecosystems and human society to global change.

The USGCRP has been strongly backed by every Administration and Congress since its inception and FY 2001 Budget Request and program plans continue the commitment to scientific research on global change that has been demonstrated in every budget submitted by the President and the Vice President. I am confident that this bipartisan tradition of support for sound science will continue.

Finally, I would like to once again commend all the program participants and the members of the Subcommittee on Global Change Research. I would particularly like to recognize the many contributions of outgoing SGCR chair Dr. Robert Corell, who stepped down this year after leading the subcommittee for a decade. Dr. Corell provided invaluable leadership and insight during his tenure, and I know that new SGCR Chair D. James Baker and Vice-Chairs Margaret Leinen and Ghassem Asrar join me in expressing our gratitude to Dr. Corell for a job well done.

Neal Lane

Neal Lane
Director

COMMITTEE ON ENVIRONMENT AND NATURAL RESOURCES

D. James Baker, Co-Chair
National Oceanic and Atmospheric Administration

Rosina Bierbaum, Co-Chair
White House Office of Science and Technology Policy

Mike Armstrong
Federal Emergency Management Agency

Ghassem Asrar
National Aeronautics and Space Administration

Kelley Brix
Department of Veterans Affairs

Eugene Conti
Department of Transportation

Jim Decker
Department of Energy

Roland G. Droitsch
Department of Labor

Delores M. Etter
Department of Defense

Terrance J. Flannery
Central Intelligence Agency

George Frampton
Council on Environmental Quality

Charles Groat
Department of the Interior

Len Hirsch
Smithsonian Institution

Kathryn J. Jackson
Tennessee Valley Authority

Eileen Kennedy
Department of Agriculture

Margaret Leinen
National Science Foundation

Paul Leonard
Department of Housing and Urban Development

Russell Moy
National Science and Technology Council

Norine Noonan
Environmental Protection Agency

Kenneth Olden
Department of Health and Human Services

David B. Sandalow
Department of State

Wesley Warren
Office of Management and Budget

Samuel Williamson
Office of the Federal Coordinator for Meteorology

Subcommittees

Air Quality
Dan Albritton (NOAA), Chair

Ecological Systems
Mary Clutter (NSF), Co-Chair
Don Scavia (NOAA), Co-Chair

Global Change
D. James Baker (NOAA), Chair
Ghassem Asrar (NASA), Vice Chair
Margaret Leinen (NSF), Vice Chair

Natural Disaster Reduction
Mike Armstrong (FEMA), Chair
John Filson (USGS), Vice Chair
Jamie Hawkins (NOAA), Vice Chair

Toxics and Risk
Norine Noonan (EPA), Chair
Kenneth Olden (HHS), Vice Chair
Sherri Goodman (DoD), Vice Chair

SUBCOMMITTEE ON GLOBAL CHANGE RESEARCH

D. James Baker, Chair
National Oceanic and Atmospheric Administration

Ghassem Asrar, Vice-Chair
National Aeronautics and Space Administration

Margaret Leinen, Vice-Chair
National Science Foundation

Margot Anderson
Department of Energy

Charles (Chip) Groat
U.S. Geological Survey

J. Michael Hall
National Oceanic and Atmospheric Administration

Jeff Miotke
Department of State

Patrick Neale
Smithsonian Institution

Aristides Patrinos
Department of Energy

Warren Piver
National Institute of Environmental Health Sciences

Fred Saalfeld
Department of Defense

Michael Slimak
Environmental Protection Agency

William T. Sommers
Department of Agriculture

Executive Office Liaisons

Rosina Bierbaum
Office of Science and Technology Policy

Peter Backlund
Office of Science and Technology Policy

Steven Isakowitz
Office of Management and Budget

Sarah Horrigan
Office of Management and Budget

Ian Bowles
Council on Environmental Quality

Cynthia Nelson
Office of the Federal Coordinator for Meteorology

TABLE OF CONTENTS

USGCRP HIGHLIGHTS AND FY 2001 BUDGET

Introduction

The U.S. Global Change Research Program (USGCRP) seeks to provide a sound scientific understanding of the human and natural forces that influence the Earth's climate system—and thus provide a sound scientific basis for national and international decisionmaking on global change issues. The USGCRP seeks to observe, understand, predict, and assess the critical natural and human-induced dynamic states and trends of the Earth's global environmental system across a wide range of time and spatial scales.

USGCRP Objectives

- Determine the origins, rates, and likely future course of natural and anthropogenic changes.
- Increase understanding of the combined effects of multiple stresses on ecosystems.
- Understand and model global environmental change and its processes on finer spatial scales and across a wide range of timescales.
- Address the potential for surprises and abrupt changes in the global environment.
- Understand and assess the impacts of global environmental change and their consequences for the United States.

This multi-agency National Research Program is coordinated through the National Science and Technology Council's Committee on Environment and Natural Resources (CENR). The CENR has established a Subcommittee on Global Change Research (SGCR) to oversee the coordination of USGCRP activities, in cooperation with the Office of Science and Technology Policy (OSTP) and the Office of Management and Budget (OMB). The SGCR is composed of representatives of the departments and agencies participating in the USGCRP, including: the Departments of Agriculture (USDA), Commerce (National Oceanic and Atmospheric Administration [DOC/NOAA]), Defense (DoD), Energy (DOE), Health and Human Services (National Institutes of Health [HHS/NIH]), the Interior (U.S. Geological Survey [DOI/USGS]), and State (DOS), the Environmental Protection Agency (EPA), the National Aeronautics and Space Administration (NASA), the National Science Foundation (NSF), and the Smithsonian Institution (SI). USGCRP science results provide useful information for environmental decision-making on issues such as climate change, stratospheric ozone depletion, changes in ecosystems, and global land cover and land use.

A Record of Accomplishment

The USGCRP began as a Presidential Initiative in 1989 and was codified by the Global Change Research Act of 1990. In the decade since, the program has led to substantial increases in knowledge, advanced predictive understanding, and documented evidence of global environmental change, including major scientific advances in the understanding of stratospheric ozone depletion, the El Niño-Southern Oscillation (ENSO) phenomenon, global climate change, tropical deforestation, and other issues.

- USGCRP-supported science helped explain the origins and behavior of the Antarctic ozone hole and showed that it was caused by human activities. Ongoing research and observations have shown that emissions controls implemented under the Montreal Protocol on Substances That Deplete the Ozone Layer have begun to decrease the concentration of several ozone-depleting gases at the Earth's surface.
- USGCRP-supported observations and analyses played a prominent role in demonstrating that emissions of greenhouse gases resulting from human activities are changing the composition of the atmosphere; that such changes have likely contributed to the global average temperature increase of between 0.7° and 1.5o F observed since about 1860; and that much larger and more rapid increases in temperature are likely to occur in the next 100 years if emissions are not reduced.
- The scientific community, working in the context of the USGCRP, successfully predicted the onset of the 1997–1998 El Niño and the subsequent La Niña, as well as some of the resulting climate anomalies around the world. Some societies were able to make limited but significant advance preparations; in some cases, economic consequences and loss of life and property were reduced.
- Satellite observations supported by the USGCRP, in coordination with cooperative international research programs, have documented and quantified large-scale changes in land cover and land use, such as the loss of tropical forest in Brazil, Southeast Asia, and Africa. Increasingly rapid changes, driven largely by human activities, are contributing to biodiversity loss, changes in atmospheric composition, and climate change.

These global environmental issues present long-term challenges at local and regional scales as well. Over the next decade, the USGCRP will support research to explore the interrelationships of these problems and identify their global, regional, and local impacts. Could global warming be influencing the timing and duration of El Niño events? Have human land-use practices—which are known to be a factor in carbon cycling—created large-scale carbon sinks, and can these sinks be maintained? How did the large, global-scale, very rapid climate changes observed in the paleoclimate record occur, and what might trigger similarly rapid changes in the future? By providing answers to such questions, science can help us better understand and deal with the causes and consequences of global environmental change.

FY 2001 Budget Highlights

For FY 2001, the President is requesting $1.74 billion for the USGCRP—an increase of $47 million above the amount enacted for FY 2000 (see Tables 1 and 2 for details). Of this amount, $845 million is for scientific research and improvements to surface-based monitoring (an increase of $87 million, or 11 percent), and $897 million is for NASA's development of Earth-observing satellites and associated data systems to help provide a fuller understanding of climate change and other global changes (a decrease of $40 million—reflecting the phasing down of funding for large development projects). Important highlights of the USGCRP budget request include the following:

- $28 million in new funding to enhance NOAA surface-based climate observations, including creation of a climate reference network to provide—for the first time—automated, simultaneous, and ideally located measurements of changing temperatures, precipitation, and soil moisture across the United States. Measurements of atmospheric trace gases, aerosols, ocean temperatures, and ocean currents also will be expanded.
- $308 million (a 13 percent increase) for research on changes in the Earth's water cycle, which is one of the primary determinants of the Earth's climate. The water cycle is emerging as a top research priority because changes appear to be occurring already. The launch of NASA's Earth Observing System (EOS) Aqua spacecraft in December 2000 will support this research by providing new global measurements of humidity, cloud properties, sea ice, precipitation, soil moisture, runoff, and snow.
- $224 million (a 9 percent increase) for research on the potential effects of climate change and other stresses on forests, coastal areas, croplands, and other ecosystems. New studies will improve our understanding of the relationships among land cover, land use, climate, and weather and assist in identifying "thresholds" for significant changes in ecosystems.
- $229 million (an 11 percent increase) for the multi-agency carbon cycle science initiative begun in FY 2000. This request includes funds to study how carbon cycles between the atmosphere, the oceans, and land and to understand and quantify the role of various natural and managed lands—including agriculture, forests, and grasslands—as sources or sinks for atmospheric carbon dioxide. Such carbon sinks may help the United States and other nations offset greenhouse gas emissions.

Table 1
U.S. Global Change Research Program

FY 1999–FY 2001 Budget by Agency
(discretionary budget authority in $millions)

AGENCY	FY 1999	FY 2000	FY 2001 Request
Scientific Research			
Department of Agriculture (USDA)	52	53	85
Department of Commerce (DOC/NOAA)	63	67	95
Department of Energy (DOE)	114	114	123
Department of Health and Human Services (HHS/NIH)	40	46	48
Department of the Interior (DOI/USGS)	27	27	25
Environmental Protection Agency (EPA)	16	21	23
National Aeronautics and Space Administration (NASA)	218	236	252
National Science Foundation (NSF)	182	187	187
Smithsonian Institution (SI)	7	7	7
Scientific Research Subtotal	**719**	**758**	**845**
Space-Based Observations			
National Aeronautics and Space Administration (NASA)	937	937	897
U.S. Global Change Research Program Total	**1,656**	**1,695**	**1,742**

Because Department of Defense (DoD) research activities are conducted for defense-related missions, they are not included in this USGCRP budget crosscut. Related DoD research does contribute to USGCRP goals, however.

DOE total for FY 2000 excludes $3.1M transferred to Small Business Innovative Research/Technology Transfer Program (SBIR/STTR).

New Research and Assessment Highlights

- USGCRP-funded analyses are revealing new information about long-term climate change. According to "proxy" temperature records embodied in glaciers and ice sheets, lake sediments, corals, tree rings, and the like, the 1990s appear to have been the warmest decade (and 1998 the warmest year) in the past 1,000 years.
- Several new analyses show that the climate record of the 20th century cannot be explained solely by accounting for solar variability, volcanic eruptions, and El Niño cycles. It appears more likely that greenhouse gases from human activities were the dominant drivers of these global-average temperature increases in the 20th century.
- New model results indicate that radiative forcing that can be attributed to tropospheric ozone since 1800 is about one-fourth the forcing from carbon dioxide (CO_2) during the same period—providing evidence that human-caused pollution in the lower atmosphere is having significant effects on global climate. This forcing is about two-thirds as large as direct forcing by tropospheric aerosols. (The latter forcing is in the opposite direction: aerosols cool, whereas ozone warms.) These forcings vary by location; sulfate aerosols have the largest effect in northern mid-latitudes, whereras ozone's contribution is largest in other locations.
- Emerging evidence suggests that we may already be witnessing the effects of climate change on terrestrial and ocean species and ecosystems. A series of scientific analyses published this year provides evidence that climate change is playing a role in the demise of several frog and toad species in Costa Rica, shifts in the growing season in Europe, shifts in growth of grass and forb species in Colorado, changes in bird ranges in western Europe, coral bleaching throughout the global tropics, and alterations in patterns of ocean biodiversity.
- The air-to-sea flux of CO_2 was measured directly for the first time using meteorological techniques on-board ship. This result paves the way for progress in understanding the processes that control the uptake of carbon by the ocean.
- Appreciable insight was gained into changes taking place at high latitudes. For example, the extent and thickness of Arctic sea ice have been decreasing, leading to significant reductions in sea ice volume. The thickness of ice cover in some regions of Greenland has been decreasing, although in other areas it is increasing. Observations in Antarctica show evidence for the existence of ice streams in the interior—the extent and rate of which exceed previously expected values.
- USGCRP-supported scientists played a key role in the Intergovernmental Panel on Climate Change (IPCC) 1999 Special Report on *Aviation and the Global Atmosphere*. This report is the first international assessment of how current and future aviation may increase global climate and stratospheric ozone depletion. The assessment was groundbreaking by virtue of its involvement of both the scientific and aviation-technology communities. It is providing key input to pending national and international decisions regarding the rapidly growing aviation industry.
- U.S. scientists also have played important roles in IPCC Special Reports on *Emissions Scenarios; Methodological and Technological Issues in Technology Transfer*; and *Land Use, Land-Use Change, and Forestry*—all of which have been

published in 2000. The USGCRP coordinates the U.S. Government's scientific and technical review of the products of these international assessments.

- A substantial number of USGCRP-supported U.S. researchers are contributing to the IPCC's Third Assessment Report on the state of understanding of climate change and its impacts, which is scheduled to be completed early in 2001.
- The USGCRP is supporting the first National Assessment of the Potential Consequences of Climate Variability and Change. The Overview report of the National Assessment Synthesis Team will be published in late 2000.

Observation, Monitoring, and Data Management Highlights

In the past year, scientists have achieved substantial scientific accomplishments in observations and monitoring to improve our understanding of the Earth system. Initial steps were taken to preserve important surface-based measurement systems for temperature and precipitation. Satellite remote sensing helped improve short-term weather prediction and quantification of the availability of fresh water globally by measuring global rainfall over the tropics; helped scientists understand the role of oceans in removing carbon dioxide from the atmosphere by producing near-daily ocean color maps; facilitated improved seasonal climate forecasts by documenting the waxing and waning of El Niño; and improved short-term weather prediction and tracking of major hurricanes and tropical storms globally by resuming global measurement of winds at the ocean surface.

Researchers used satellite radar data to determine thinning and thickening rates for the Greenland ice sheet, provide the first detailed radar mosaic of Antarctica, and provide daily observations of the polar regions from space. They produced the first satellite-derived assessments of global forest cover, continued to measure concentrations of ozone and ozone-depleting substances, and implemented a 17-year data record of aerosols and cloud properties directed toward predicting annual to decadal climate variations. Finally, a series of new satellites—including the first elements of the NASA Earth Observing System (EOS), which has been under development for the past decade—was launched successfully.

- EOS-Terra (previously called EOS-AM), the flagship of the EOS satellite series, was launched in December 1999. Terra will observe clouds, aerosols, trace gases, land surface, and ocean properties, as well as the Earth's radiation budget. It will make important contributions to USGCRP investigations of the carbon cycle, climate change, atmospheric chemistry, ecological changes, and the water cycle.
- The Active Cavity Radiometer Irradiance Monitor (ACRIMSAT) satellite was launched in December 1999 to consolidate and extend more than 20 years of observations of total solar irradiance. Continuity in satellite-based solar observations is essential for USGCRP climate change research to elucidate solar-terrestrial connections and the effect of solar variations on the atmosphere and weather, and to distinguish between natural variability caused by solar forcing and that induced by anthro-

pogenic greenhouse gases.

- Landsat-7 was launched in April 1999. High-quality data distribution began in August 1999. Seasonal image collection to refresh the global archive began in July 1999, and more than 50,000 acquisitions were archived. Landsat-7 also includes a 15 m resolution panchromatic band for the study of ecosystem disturbances. This imaging contributes directly to the monitoring of land cover and land-use changes—key elements for understanding and quantifying how such changes affect the global and regional carbon cycle and the biology and biogeochemistry of ecosystems.
- The QuikSCAT spacecraft, launched in June 1999, joins the Tropical Rainfall Measuring Mission (TRMM) and the Ocean Topography Experiment/Poseidon (TOPEX/Poseidon) to form a powerful suite of space-based observational assets to track phenomena such as El Niño/La Niña events. QuikSCAT measures sea-surface wind speed and direction at a spatial resolution of 25 km over at least 90 percent of the ice-free global oceans every two days. Data from this mission are being used to improve short-term weather forecasts and are important for climate change research.
- Through the EOS Data and Information Systems (EOSDIS), Earth science data products are provided routinely to end-users within five days of receipt or production of the requested data product. These products comprise data from currently operating space assets, including: precipitation measurements and observations of tropical storms from TRMM, ocean productivity measurements from the Sea-viewing Wide Field-of-view Sensor (SeaWiFS), detection of ocean surface height changes used to predict El Niño occurrence and strength from TOPEX/Poseidon, and sea-ice motion and Antarctic mapping from Canada's RADARSAT. The data also include measurements of stratospheric trace chemicals from the Upper Atmospheric Research Satellite (UARS), Antarctic ozone hole measurements from the Total Ozone Mapping System (TOMS), land use and land cover data from heritage Landsat missions, and measurements of incoming solar radiation and outgoing radiation from the Earth by the Earth Radiation Budget Experiment (ERBE). Similar arrangements are in place to meet the anticipated demand for data products from EOS-Terra, EOS-Aqua, and other satellite missions.

The National Assessment of the Potential Consequences of Climate Variability and Change

It has long been known that climate varies, and that such variations can have important effects on our society. One of the things we have learned over the past few decades is that we are changing the concentrations of greenhouse gases, some of which have very long atmospheric lifetimes. The consequent effects also will be long-lived. Past emissions of greenhouse gases and the long-term nature of the atmosphere's response already have committed us to a degree or more of additional warming. Although reducing emissions clearly will slow the increase in atmospheric concentrations and reduce the amount and rate of climate change in the future, some further change will be inevitable because of emissions that have already occurred. As these changes are occurring, we will continue to experience climate variations, with some types of events occurring more often and some less often.

The National Assessment effort now underway in the USGCRP is examining the degree to which particular regions and sectors of the United States are vulnerable to climate variations and change, the potential ecological and socioeconomic impacts of climate variations and change, and options available to adapt and prepare for the next few decades and the next century. The National Assessment also is identifying key information gaps and research needs (i.e., information that is still required to answer questions of interest to resource managers and decisionmakers).

The assessment effort has included a series of regional workshops with participation by a broad range of public and private stakeholders, to identify issues of interest, as well as a series of regional and sectoral analyses, most of which are being completed this year. The national-level product of the assessment is an Overview report by the National Assessment Synthesis Team that is expected to be completed in fall 2000. The Overview is undergoing a rigorous peer review that includes several rounds of technical review, full agency review, and a 60-day public comment period before it is submitted to the President and the Congress. The Global Change Research Act of 1990 calls for this type of assessment of the potential consequences of global changes on a periodic basis (not less than every four years). Assessment will continue to be an important ongoing component of the USGCRP.

Although the first National Assessment will soon be completed, we expect many of the lessons learned during this process to play a significant role in the definition of future USGCRP research activities. This first assessment is part of a larger evolution of the USGCRP. During much of the first decade of its existence, the USGCRP concentrated on observing and documenting change in the Earth's physical systems and understanding why these changes are occurring. The USGCRP is now appropriately developing a broader research effort that also includes improved understanding of how global change will affect the Earth's biological systems—and the human societies that are dependent upon them—and making useful scientific data and information more broadly available for public and private planning and decisionmaking.

To accomplish these goals, we must greatly improve our capabilities for conducting regional-scale assessment of global change and its potential consequences around the country. Our current level of understanding tells us that climate change and its effects will vary by region, but our ability to project specific regional effects remains limited.

We also need to learn more about the interactions of natural and human-induced climate change and variability and other human-induced changes in the environment—such as pollution, land-use change, resource extraction, and invasive species, many of which are regional in scale. In addition, we need to achieve an integrated understanding not only of the nature and extent of physical and biological effects of climate change but also of their ramifications for our social and economic systems.

Table 2
U.S. Global Change Research Program

FY 2000 – FY 2001 Budget by Program Element
(discretionary budget authority in $millions)

PROGRAM ELEMENT	FY 2000	FY 2001 Request
Understanding the Climate System	498.6	485.4
Understanding the Composition and Chemistry of the Atmosphere	387.9	365.4
Global Water Cycle	273.8	308.1
Global Carbon Cycle	204.2	229.2
Understanding Changes in Ecosystems	205.6	223.6
Understanding the Human Dimensions of Global Change	91.8	95.3
Paleoclimate: The History of the Earth System	29.6	27.4
U.S. Global Change Research Total	**1,691.4**	**1,734.7**

USGCRP total excludes $4.0 million for DOI/USGS data management in FY 2000 and FY 2001 and $3.1 million for DOE Small Business Innovative Research/ Technology Transfer (SBIR/STTR) program in FY 2000 and FY 2001.

Because DoD research activities are conducted for defense-related missions, they are not included in this USGCRP budget crosscut. Related DoD research does contribute to USGCRP goals, however.

NEAR-TERM PLANS

Understanding the Earth's Climate System

The USGCRP budget includes $485 million in FY 2001 for programs related to understanding the Earth's climate system (see Table 3). Climate is a naturally varying and dynamic system with important implications for the social and economic well-being of our societies. Understanding and predicting climate changes across multiple timescales (from seasonal to interannual to decadal and longer) offer valuable information for decisionmaking in sectors that are sensitive to rainfall and temperature fluctuations, such as agriculture, water management, energy, transportation, and human health.

A recent report from the National Research Council (NRC) offers several important conclusions related to long-term changes in surface temperature. The NRC report concluded that the surface of the Earth has warmed over the past 120 years and that the warming trend during the past 20 years is substantially greater than the average rate of warming during the 20th century. The NRC also stated that "the disparity between surface and upper air trends in no way invalidates the conclusion that surface temperature has been rising." Additional recent evidence of decreases in the extent of Arctic sea ice suggests that natural variability alone is not sufficient to explain the warming.

A fundamental challenge that follows from these analyses is determining how much of the observed change in the climate is attributable to human activities and how much to natural variability. Responding to this challenge requires improving our understanding of natural variability and human influences. Such improvement requires a balance of observations, studies of underlying Earth system processes, and predictive modeling.

The USGCRP conducts climate research within an "end-to-end" framework. The end-to-end process transforms integrated process research (such as understanding ENSO and other modes of variability), data and observations, and modeling activities into useful information and prediction products that decisionmakers can understand and use. The end-to-end process also entails an understanding of the relationship between humans and climate (including impacts) and a purposeful and focused interaction with the users of USGCRP research to reconcile research directions with societal needs. As part of this endeavor, seasonal to interannual forecasts are compared to actual observations, and models are improved accordingly. Over time, the end-to-end approach has led to a significant improvement in shorter-term climate predictions and, because of their useful accuracy, their use in risk management. (Much of the research on understanding the relationship between humans and climate is supported as part of the "Understanding the Human Dimensions of Global Change" USGCRP Program Element.)

Recent Accomplishments

- New research findings have identified declines in the extent of Arctic sea ice and its thickness over the past several decades. The average thickness from the ice surface to the bottom of the ice pack has declined by about 40 percent. A related study used climate models to estimate that the probability that the observed trends could be caused entirely by natural variability is less than 2 percent. This research suggests that human activities are very likely contributing to the loss of Arctic sea ice.
- Scientists have showed that sea-surface temperature changes associated with the Pacific Decadal Oscillation (PDO) have a significant effect on the steering of ocean storms across the North Pacific Ocean, thereby influencing North American weather. There is evidence that reversals in the PDO phase occurred around 1925, 1947, and 1977. Because the PDO tends to persist in one phase or the other for multi-year periods, being able to account for its phase in the future is potentially important in increasing the accuracy of seasonal to interannual forecasts.
- Research results demonstrate increases in the seasonal predictability of U.S. precipitation and surface air temperature when the combined impacts of ENSO, the PDO, and the Arctic Oscillation (AO) are all taken into account. Considering ENSO and the PDO together enhance predictive accuracy for precipitation, primarily across the southern states; adding the AO adds significantly to the predictive accuracy over the eastern states.
- Observations and models have demonstrated a link between decadal variations in the production of Labrador Sea water and large-scale surface temperature anomalies in the tropical Atlantic Ocean. This finding suggests that there is a link between climate variability at high and low latitudes and that the deep ocean may influence climate variability on decadal timescales.

Table 3
Understanding the Climate System

FY 2001 Budget by Agency
(discretionary budget authority in $millions)

DOC/NOAA	Scientific Research	41.1
DOC/NOAA	Surface-Based Observations	19.8
DOE		69.5
NASA	Scientific Research	62.3
NASA	Space-Based Observations	208.8
NSF		83.5
Smithsonian Institution		0.4
TOTAL		**485.4**

Observed Hemispheric Asymmetry in Global Sea Ice Changes

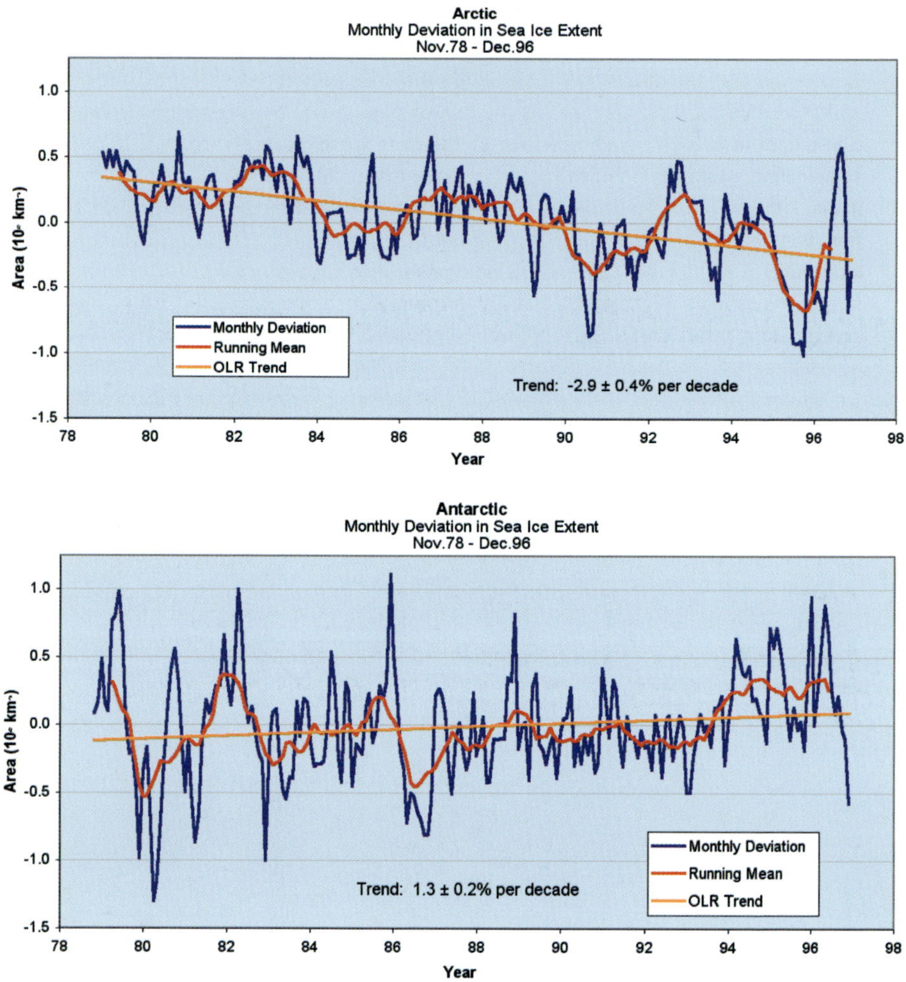

Figure 1. Asymmetric Trends in Arctic and Antarctic Sea-Ice Extent

(See page 38 for additional information)

FY 2001 Plans

The USGCRP will improve its surface-based climate observation systems to provide more accurate, timely measurements in the United States and more extensive observations of the global oceans. These steps are part of a multi-year initiative to build and sustain global climate and ocean observing systems that meet the long-term observational requirements of the USGCRP, operational forecast centers, and major scientific assessments. Goals include:

- Increasing the accuracy of annual total surface-based precipitation measurements by 5-30 percent, by installing and beginning to provide data from the first 100 sites of a 250-station Climate Reference Network across the United States (currently, during mid-winter months, precipitation measurements in some locations are only 25 per cent accurate; the new stations should increase this to 50-75 percent). Each automated site will make long-term, bias-free measurements of temperature, precipitation, wind speed, and soil moisture. For the first time, a network outside the urban environment will be available to calculate precisely the number of days that exceed specific thresholds of temperature and precipitation—for example, 100 mm/day of precipitation or 30° C (the temperature above which corn growth is severely constrained).
- Extending the coverage of sea-surface buoy networks from 75 percent to about 85 percent of what is required to produce a weekly global sea-surface temperature analysis, in combination with satellite remote-sensing measurements, on a 5° grid with errors below 0.5° C. Extend upper-ocean network coverage from a current state of about 25 percent to about 30 percent of what is required to produce a weekly upper-ocean analysis, in combination with satellite measurements, on a 3° grid to 0.5° C accuracy. These improvements will result from deployment of additional drifting and moored buoys, complemented by additional high-resolution volunteer-observing-ship lines for measuring air-sea fluxes and upper-ocean structure.

The USGCRP will strengthen the ability to simulate the Earth's climate, by improving climate model resolution, increasing the ability to conduct long-term climate simulations, and testing improved calculations of climate processes. Goals include:

- Improving model resolution by 100 percent with a new version of the community Climate System Model (CSM-2), which will enable inclusion of smaller-scale ocean processes and better representation of mountains and land-surface energy and water exchanges. The USGCRP will complete a 500-year simulation of the large-scale features of the global climate system and produce a report evaluating the ability of CSM-2 to simulate natural climate variability on seasonal to centennial timescales.
- Developing more scalable code for the atmospheric and flux coupler components of a coupled climate model to enable effective exploitation of several hundred processors (a factor of two to three times more than at present) for a single simulation at moderate resolution. This is required to conduct the series of long-term (century or longer) simulations using high-end parallel architecture computers that will be critical in determining the potential usefulness of next-generation high-end computers for production-level climate model runs.

- Using a detailed cloud climatology derived from three years of continuous observations of clouds and radiation in the Southern Great Plains to improve processes for calculating the radiative effects of clouds on climate. The USGCRP will test the improvements using cloud-resolving models and mesoscale models and report on the percentage improvement in accuracy achieved.

The USGCRP will continue to expand its space-based observation system through the successful launch and operation of NASA Earth Observing System (EOS) and Earth System Science Pathfinder (ESSP) satellites. Goals include:

- Successful collection and analysis of simultaneous measurements of clouds, aerosols, trace gases, land surface and ocean properties, and the Earth's radiation budget from the EOS-Terra satellite (launched in FY 2000). Terra will provide, for the first time, simultaneous measurements of critical parameters of the whole-Earth system, as well as integrated analysis of land-ocean-atmosphere interactions.
- Successful launch of the EOS-Ice-Sheet Altimetry Mission (IceSat) and initiation of measurements to establish ice sheet mass balance and cloud-top and land surface topography, vertical profiles of aerosols, and cloud properties. IceSat will provide the first viable measurements of surface elevations for the high-relief margins of the ice sheets.
- Successful launch of the second ESSP mission—the Gravity Recovery and Climate Experiment (GRACE)—and initiation of measurements of the Earth's gravity field. Data from GRACE will enable scientists to infer the movements of ocean currents, changes in ocean mass, and variations in the size of ice sheets and glaciers and monitor soil moisture and the state of major aquifers. When data from GRACE are combined with other data, more ocean properties can be retrieved—for example, allowing thermal expansion of the upper ocean to be distinguished from mass changes caused by the addition of water from sources such as melting glaciers.
- Creating a near-global high-resolution digital elevation topographic map of the world with about 1000 times greater spatial resolution than the existing global map by completing the analysis and processing of data from the Shuttle Radar Topography Mission (SRTM) instrument. This improved resolution will support research and applications in ecology, geology, geodynamics, hydrology, and atmospheric modeling, as well as applications such as urban and infrastructure planning and disaster management.

Understanding the Composition and Chemistry of the Atmosphere

The USGCRP budget includes $365 million in FY 2001 for programs to study the composition and chemistry of the atmosphere (see Table 4). Changes in the global atmosphere can have important implications for life on Earth, including such factors as the exposure to biologically damaging ultraviolet (UV) radiation, the abundance of greenhouse gases and aerosols (which, in turn, affect climate), and regional air pollution.Thus, this research is a central component of our effort to understand global change. This area of research has a strong record of accomplishment and provides good examples of how research can establish the scientific basis for effective policymaking.

For example, reductions in the total amount of stratospheric ozone over most of the Earth have been observed during the past 20 years. A combination of aircraft-, ground-, balloon-, and satellite-borne instruments have shown that industrially produced chlorine- and bromine-containing compounds are primarily responsible for the observed ozone depletion. As the Montreal Protocol on Substances That Deplete the Ozone Layer has been implemented, atmospheric concentrations of several of the regulated long-lived compounds have stopped increasing and have begun to decline. Scientists believe that maximum levels of stratospheric chlorine have been reached, and evidence suggests that they have begun to decline. This decrease should lead to a healing of the ozone layer in about 50 years, although unquantified effects of climate change on ozone chemistry may significantly affect the nature and timing of expected ozone recovery.

Human activity that can affect atmospheric composition includes the use of chlorofluorocarbons (CFCs) and other halogenated hydrocarbons, fossil fuel combustion and the associated release of air pollutants, and changes in agricultural practices that affect the concentration of gases such as nitrous oxide and methane, as well as that of smoke. Scientists believe that these changes have led to significant changes in the distribution of ozone in the troposphere; this ozone is both an air pollutant and a contributor to climate change. Chemical processes in the troposphere are important in converting industrial emissions into aerosol particles, which have effects on climate and on the tropospheric chemistry. Changes in climate that are driven largely by increases in greenhouse gases and aerosols of anthropogenic origin also can be expected to affect atmospheric chemistry in complex ways that are difficult to predict. Natural processes, such as volcanic eruptions and variations in solar output, also can affect the chemistry of the atmosphere.

Recent Accomplishments

- New observational and model results have shown that the chemistry and meteorology of the stratosphere are influencing the dynamics of the troposphere, including tropospheric temperatures. Thin layers in the troposphere appear to form from the lateral and downward transport of air from the stratosphere and the lofting of pollution into the middle and upper troposphere.
- Long-term studies of the polar vortices that trap air throughout the winter and enhance ozone depletion show that, since the 1980s, there has been an increase in the persistence of these patterns, with larger variations in the Arctic than in the Antarctic.
- Aircraft measurements carried out as part of the Stratospheric Aerosol and Gas Experiment (SAGE III) Ozone Loss and Validation Experiment (SOLVE) campaign have provided new information about the formation of polar stratospheric clouds and their effects on atmospheric chemistry in the Arctic region. Detailed analyses of SOLVE data should reduce uncertainties about the possibility of continuing Arctic ozone loss over the next decade and the nature and timing of the expected long-term ozone recovery.

Polar ozone loss is now evident in both hemispheres, raising concern that climate change could delay the recovery of stratospheric ozone

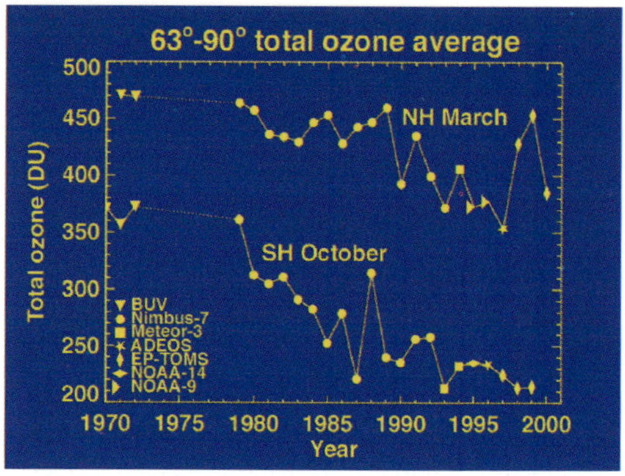

Figure 2a. Polar Ozone Loss

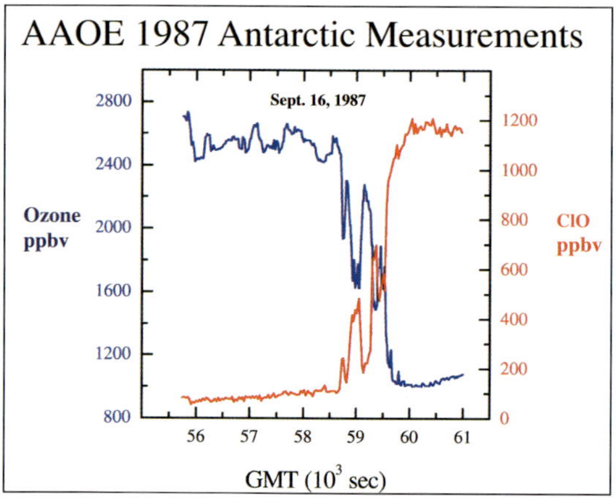

Stratospheric ozone is depleted, as the level of active chlorine (ClO) rises,along the flight track of NASA's ER-2 high-altitude airborne laboratory

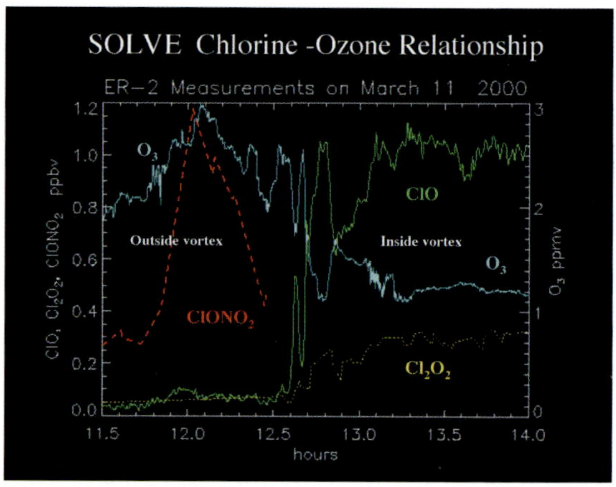

Figure 2b and 2c.
Stratospheric Ozone Depletion

(See pages 38-39 for additional information)

- The Indian Ocean Experiment (INDOEX)—an international study of how air pollutants are transported through the atmosphere and how they affect atmospheric composition and solar radiation processes over the ocean—is finding that particles of soot produced in southern Asia can be a significant contributor to surface warming in the Indian Ocean. These pollutants heat the lower atmosphere directly by absorbing sunlight and indirectly by burning off clouds and thus may have important consequences for the region's climate and hydrologic cycle.
- The Pacific Exploratory Mission in the Tropics provided the first extensive measurements of the hydroxyl radical (OH) oxidant in the tropical troposphere. These measurements demonstrated that global model predictions of atmospheric OH levels are basically correct. These model predictions have played a critical role in environmental policy—notably for the replacement of CFCs by products that undergo oxidation by OH in the troposphere—and their verification represents a critical milestone for atmospheric chemistry.

Table 4
Understanding the Composition and Chemistry of the Atmsophere

FY 2001 Budget by Agency
(discretionary budget authority in $millions)

USDA		17.6
DOC/NOAA	Scientific Research	8.8
DOC/NOAA	Surface-Based Observations	1.0
DOE		12.6
NASA	Scientific Research	53.7
NASA	Space-Based Observations	252.7
NSF		18.7
Smithsonian Institution		0.3
TOTAL		**365.4**

FY 2001 Plans

The USGCRP will improve its atmospheric chemistry measurement capabilities. Goals include:

- Upgrading and expanding the operations of the Global Atmospheric Baseline Observatories at Barrow, Alaska; Mauna Loa, Hawaii; American Samoa; and the South Pole.
- Extending routine aircraft flask profile measurements of trace gases (carbon dioxide,

methane, carbon monoxide, and possibly some CFCs) and aerosols to cover the continental United States and the Pacific Basin.

- Doubling the number of U.S. sites (from two to four) where measurements are taken above the surface by aircraft.
- Establishing baseline observations of surface ozone, carbon monoxide, aerosols, and possibly sulfur dioxide at a U.S. West Coast site, against which future data can be compared. (Intercontinental transport of Asian pollution is expected to be a factor in the future).
- Successful launch of the Quick Total Ozone Mapping Spectrometer (QuikTOMS) spacecraft and the SAGE III instrument, initiating measurements of atmospheric ozone from both missions, and making the data products available to the public. This will extend the long-term record of atmospheric ozone concentrations that is critical for atmospheric chemistry research and continued monitoring of the progress achieved as a result of the Montreal Protocol. The data should help in documenting the continued vulnerability of the ozone layer during the next decade as stratospheric chlorine levels slowly decline.
- Producing a summary of long-term, highly calibrated observations of surface UV flux and the potential ramifications of changes from the USGCRP interagency ground-based UV radiation network, and providing this data to users in the physical sciences, ecology, and human health communities.

The USGCRP will also conduct a wide range of research activities aimed at improving our understanding of global and regional atmospheric processes. Goals include:

- Improving our understanding of tropospheric ozone and the other components of the troposphere that regulate it by sponsoring individual studies and broad measurement campaigns.
- Improving our understanding of atmospheric processes as well as the impact of ozone and aerosols on radiative forcing in the East Asia/western Pacific Ocean region by conducting field campaigns (TRACE-P and ACE-Asia) to obtain regional and process-scale data. This will provide an important early picture of the atmosphere in this region that will help us understand the changes we expect to see in the next few decades.
- Producing the first annual global climatology of the three-dimensional structure of carbon monoxide and the spatial dependence of total column methane in the troposphere, using data obtained from the Measurements of Pollution in the Troposphere (MOPITT) instrument on NASA's EOS-Terra spacecraft (launched in FY 2000). The USGCRP will provide this information to researchers and interested members of the public through EOSDIS's Distributed Active Archive Centers, which are accessible through the USGCRP Web site.

The Global Water Cycle

The USGCRP budget includes $308 million in FY 2001 for the study of the global water cycle (see Table 5). The cycle of water through the land, atmosphere, and oceans is intimately tied to the Earth's climate through processes such as latent heat exchange and the radiative effects of water in its vapor, liquid, and solid phases. Water, and its cycling in the Earth system, is critical for human populations and ecosystems. The National Climate Assessment process is clearly identifying changes in the timing and availability of water as central to an understanding of the effects of climate change.

Projections of human demand for and availability of fresh water suggest that the world may face severe limits in the next 50 years. Water resources are used and managed primarily at the local scale, but water cycle processes that are responsible for sustaining and renewing them operate over scales as large as the globe. Fluctuations in these water cycle processes induce severe weather and hydrologic extremes, such as droughts and floods, that affect economic infrastructures, human health, and ecosystems.

The study of the global water cycle is a unifying theme that bridges the gap between the spatial scales involved in global atmospheric (and atmosphere-ocean interaction) processes and land surface hydrological processes—both of which determine the availability of water resources. The water cycle is emerging as a top research priority because changes in the cycle appear to be occurring already. The primary goals of this research are a greater understanding of the variability of the hydrologic cycle on seasonal, annual, and interannual timescales and on regional to global spatial scales, as well as improved ability to predict or anticipate short- and longer-term changes—and thus a greater understanding of the hydrological interactions among the terrestrial, atmospheric, and oceanic components of the Earth's climate system.

Recent Activities

In FY 1999, the USGCRP established a Water Cycle Study Panel and charged it with redefining the USGCRP approach to the study of the water cycle. This group, which includes government and academic scientists, is developing a new set of comprehensive research and applications strategies that will take advantage of existing and planned observing systems to address major issues concerning the global water cycle and global and national water resources. A combination of observations, modeling, and analysis at a range of spatial and temporal scales will provide the foundation for understanding the water cycle. The ultimate goal of the water cycle research program is to address three fundamental scientific questions:

- What are the underlying causes of variations in the water cycle on global and regional scales, and to what extent is this variation induced by human activity?
- To what extent are variations in the global and regional water cycle predictable?
- How will variability and changes in the cycling of water through terrestrial and freshwater ecosystems be linked to variability and changes in the cycling of carbon, nitrogen, and other nutrients at regional and global scales?

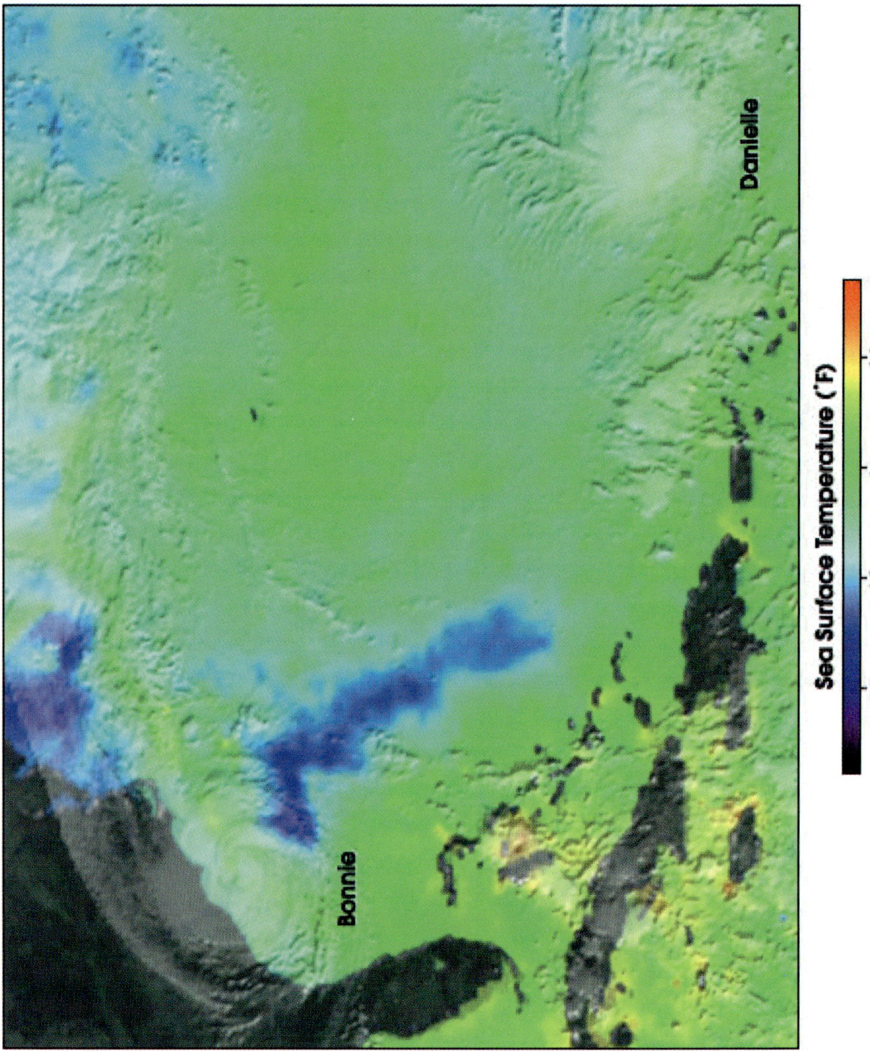

Figure 3. TRMM Measures Sea Surface Temperature Through Clouds

(See pages 39-40 for additional information)

Table 5
The Global Water Cycle

FY 2001 Budget by Agency
(discretionary budget authority in $millions)

USDA		0.3
DOC/NOAA	Scientific Research	5.0
DOC/NOAA	Surface-Based Observations	1.9
DOE		3.0
NASA	Scientific Research	59.5
NASA	Space-Based Observations	228.7
NSF		9.7
TOTAL		**308.1**

FY 2001 Plans

The USGCRP, in collaboration with the science community, will complete and publish a comprehensive water cycle science plan that will define key research questions and set new near-term and longer-term observational, modeling, and research priorities. The USGCRP will also continue improving its capability to measure important aspects of the global water cycle and will conduct a number of important field campaigns and research projects. Goals include:

- Reducing the uncertainty in global tropical rainfall estimates to about 10 percent, an 80 percent improvement over the current uncertainty of about 50 percent, by completing the fourth year of instantaneous measurements of global tropical rainfall rates and accumulation in the global tropics using the NASA TRMM satellite.
- Successful launch of the NASA EOS-Aqua satellite (scheduled for December 2000) and initiation of the first simultaneous global measurements of atmospheric temperature and water vapor profiles, radiation budget, clouds, precipitation, sea-surface winds and temperature (which determine evaporation and latent and sensible heat fluxes), vegetation cover (a determining factor in evapotranspiration), ice, snow, and surface wetness. These data will support a series of focused investigations of various elements of the global water cycle.
- Completing two field campaigns (and follow-on data analysis) at the Atmospheric Radiation Measurement Southern Great Plains Site. One of these campaigns will establish the accuracy of several commonly used techniques for obtaining water vapor profiles; the other will explore changes in the distribution of water vapor over the site. Both will contribute to increased understanding of the dynamics of the water cycle—in particular the distribution and rates of change of water in the atmosphere.
- Reducing uncertainty about the movement of nutrients from land through coastal

wetlands into the near-shore bay environments through support of integrated research projects on coastal ecosystems—focusing on three new land-coastal margin Long-Term Ecological Research sites added to the network in FY 2000. These long-term collaborative research projects will integrate research on hydrology, biogeochemistry, and ecology.

- Improving the capability to predict changes in the hydrological cycle and water resources on seasonal timescales through initiation of the America Prediction Project as part of the Global Energy and Water Cycle Experiment (GEWEX). This project will place greater emphasis on prediction than has been the case with earlier diagnostics-oriented research. It will cover all of the United States, with a special emphasis on land-atmosphere interactions in the West—including the role of land in monsoonal circulations and the role of mountains in the hydrologic cycle.

The Global Carbon Cycle

The USGCRP budget includes $229 million in FY 2001 to continue the Carbon Cycle Science Initiative that began in FY 2000 (see Table 6). This ongoing effort will provide critical scientific information on the fate of carbon dioxide emitted to the atmosphere, sources and sinks of carbon on continental and regional scales, and how sinks might change naturally over time or be modified by agricultural or forestry practices. USDA, DOE, DOI/USGS, NASA, NSF, DOC/NOAA, and the Smithsonian Institution will all play important roles in this effort, guided by a science plan that has been drafted with participation by many of the leading scientists in this field.

The Carbon Cycle Science Initiative will employ a wide variety of research activities in a comprehensive examination of the integrated global carbon cycle, with an initial emphasis on North America. Comparison of North America with other regions also will be important for understanding the relative importance of our region in the global context. Atmospheric and oceanographic field sampling campaigns over the continent and adjacent ocean basins will be combined with atmospheric transport models to develop more robust estimates of the continental distribution and subcontinental-scale magnitude of North American carbon sinks. Local-scale experiments conducted in various regions will contribute to the understanding of mechanisms involved in the operation of carbon sinks on land and in the ocean, the quantities of carbon assimilated by ecosystems, and how these quantities might change in the future.

The initiative also will include evaluation of information from past and current land-use changes—from remotely sensed and historical records—to assess how human activity has affected carbon storage on land. Potential management strategies for maximizing carbon storage will be studied, including evaluation of the variability, sustainability, lifetime, and related uncertainties of different managed sequestration approaches. Finally, enhanced long-term monitoring of the atmosphere, ocean, forests, agricultural lands, and rangelands, using improved inventory techniques and new remote-sensing capabilities, will be initiated to determine long-term changes in carbon stocks. Integration of new observations and understanding of carbon cycle processes into regional and global carbon system models will enable us to project more accurately future atmospheric concentrations of carbon dioxide and other greenhouse gases. The ultimate goal of the carbon cycle science program is to answer the following fundamental questions:

- What has happened to carbon that has already been emitted by human activities?
- What will be the future atmospheric carbon dioxide concentration resulting from past and future emissions?
- How do land management, land use, and other factors affect carbon sources and sinks over time?
- How will future environmental changes and human actions affect atmospheric concentrations of carbon-containing greenhouse gases?

Recent Accomplishments

- Net carbon uptake by terrestrial ecosystems was determined from the AmeriFlux Network, which produces unique measurements of the net annual exchange of CO_2 between the atmosphere and terrestrial ecosystems. Data from 12 locations show that net carbon uptake is greater in warmer zones along the north-to-south climatic gradient from Canada to the southeastern United States. These observations are at variance with the conventional wisdom that more carbon accumulates at higher latitudes under colder temperatures. Scientists consider these carbon gains by the terrestrial biosphere significant.
- Three ecosystem models simulating the impact of increasing CO_2 and climate on net carbon storage in U.S. terrestrial ecosystems—and agreeing within 25 percent—have yielded estimates of a land carbon sink that corresponds to about one-third of the estimated total carbon sink, based on inventory data. These model results suggest that other processes, such as regrowth on abandoned agricultural land or harvested forest lands, have larger effects on carbon storage and highlight the need for data on land-use history and more integrated modeling approaches. The model results also show evidence of significant year-to-year variability in carbon storage; variations of 100 percent from year to year are attributable to climate variability.
- USGCRP agencies successfully implemented projects in Iowa and Montana to encourage changes in land management that should lead to increased carbon storage.
- Long-term field experiments in which CO_2, water, and nutrients have been manipulated are producing unique data on ecosystem response to these global change variables. Results include increased vegetation growth, changes in water use, and increased carbon gain by several woody and herbaceous ecosystems. The observed decline in nitrogen content in plant tissues has implications for the quality of forage for animals that might graze in these systems, however.
- A comprehensive synthesis effort is providing an inventory of carbon storage in the world's oceans that is based on observations. Previous estimates of the ocean sink had relied solely on model simulations. For example, the synthesis has revealed that more than 20 billion metric tons of excess atmospheric CO_2 are stored in the Indian Ocean. This effort represents an order of magnitude increase in the quantity and quality of carbon data obtained for the ocean as a result of improved analytical techniques and standards.

Table 6
The Global Carbon Cycle

FY 2001 Budget by Agency
(discretionary budget authority in $millions)

USDA		37.2
DOC/NOAA	Scientific Research	4.6
DOC/NOAA	Surface-Based Observations	5.3
DOE		15.6
DOI/USGS		3.5
NASA	Scientific Research	45.0
NASA	Space-Based Observations	104.6
NSF		13.1
Smithsonian Institution		0.3
TOTAL		**229.2**

FY 2001 Plans

The highest priorities for FY 2001 will continue to be understanding and quantifying North American carbon sources and sinks and filling critical gaps in our understanding of the causes of carbon sinks on land as well as processes controlling the uptake and storage of carbon in the ocean. Research advances on these issues will provide information needed as a basis for sound policymaking, as well as valuable information about potential management strategies to land and forest managers in the public and private sectors. Goals include:

- Achieving more comprehensive CO_2-exchange data from key biotic and climatic zones across the United States by adding measurements at new and existing sites in the AmeriFlux research network. Selection of additional sites will be based on competitively reviewed recommendations from the scientific community.
- Reducing uncertainty about carbon gains and losses by the terrestrial biosphere by obtaining additional data on ecosystem processes that control the exchange of CO_2 with the atmosphere. This will include simultaneous measurements of atmospheric CO_2 changes, use of isotopic methods, and other approaches. These studies will be an important component of investigations of the North American terrestrial carbon sink.

The USGCRP also will improve observational capabilities and develop new techniques to measure carbon stored in forests, soils, and agricultural lands. New satellites will obtain systematic global coverage of the pattern of aboveground vegetation and its photosynthetic carbon uptake. Goals include:

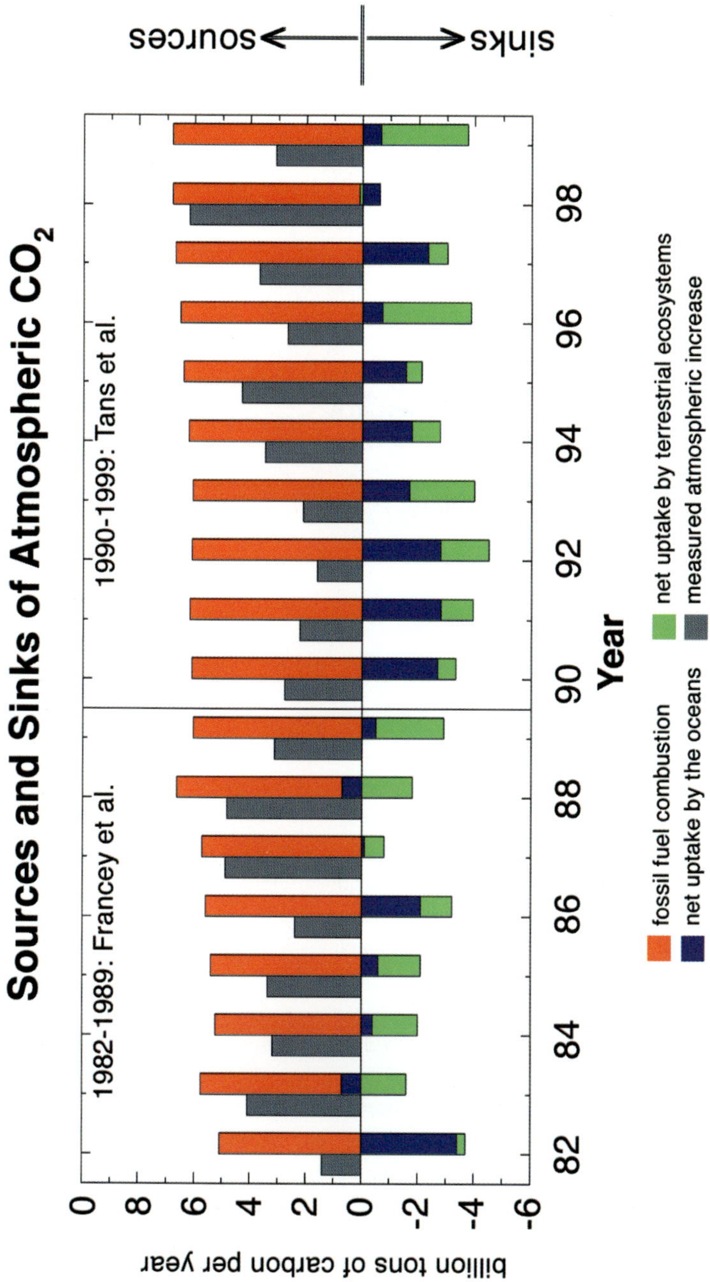

Figure 4. Partitioning of Fossil-Fuel-Derived Carbon

(See page 40 for additional information)

- Generating the first quantitative, geo-referenced, spatially comprehensive estimates of global terrestrial net primary productivity using data from the Moderate Resolution Imaging Spectroradiometer (MODIS) instrument on NASA's EOS-Terra satellite.
- Producing a quantitative assessment of which approaches for measuring carbon work best for a given scale. The USGCRP will acquire, compare, and integrate measurements at local, regional, and global scales, and reconcile differences among measurements and modeling approaches.
- Reducing the uncertainty of estimates of exchange of CO_2 between the ocean and the atmosphere in the North Atlantic and Pacific to less than a factor of two. This improved understanding of the role of the ocean in the carbon cycle will help to constrain estimates of the Northern Hemisphere terrestrial sink.

Understanding Changes in Ecosystems

The USGCRP budget includes $224 million in FY 2001 for the study of changes in managed and unmanaged ecosystems (see Table 7). The biosphere consists of diverse ecosystems that vary widely in complexity and productivity, in the extent to which they are managed, and in their economic value to society. Ecosystems directly provide forage, timber, fish, food, and fiber, as well as other services such as water cycling, climate regulation, recreational opportunities, and wildlife habitat. Ecosystems respond to and contribute to global change.

On one hand, changes in temperature and precipitation may alter the ranges of plant and animal species. The rates of change projected over the next century are faster than any that have occurred in the past 10,000 years; therefore, the character and diversity of ecosystems could undergo substantial change if their constituent species cannot adapt. Ecosystems also contribute to global change, however, by playing a role in modifying the atmosphere—and hence the climate system. Terrestrial and marine ecosystems play a role in carbon storage and the release of carbon to the atmosphere; they also are sources and sinks of other greenhouse gases, including methane and nitrous oxide. In addition, vegetation and soils influence climate by affecting the amount of radiation reflected or absorbed, the evapotranspiration of water, and other feedbacks to temperature, precipitation, and weather systems.

Management of ecosystems and natural resources will be an important aspect of society's response to global change. Ecosystems have the capacity to respond to stress; when that capacity is exceeded, however, natural resources and services are altered and begin to decline. Our ability to achieve a sustainable future depends on the protection of public lands and other lands, sustainable use of terrestrial and aquatic renewable resources, and more efficient use of nonrenewable resources. Better scientific understanding of the processes that regulate ecosystems and the capability to predict ecosystem changes and evaluate the potential consequences of management strategies will improve our ability to manage for sustainability.

Recent Accomplishments

- Research has documented significant changes in the growth and development of Ponderosa and Jeffrey pine in response to elevated ozone exposure and elevated nitrogen deposition. Above-ground biomass increases, while root biomass decreases with exposure to ozone and nitrogen, raising important questions about predisposing trees to drought-induced mortality and other stressors. The interaction of ozone and nitrogen pollution has significant implications for the storage of carbon in soils, forest-floor litter, and woody biomass.
- A new assessment of fire risk from climate change uses results from Mapped Atmospheric-Plant-Soil System (MAPSS) vegetation distribution model simulations for seven future climate scenarios. The dynamic simulations indicate that climate change could lead to increased fire frequency over much of the western United States and, under scenarios that project the greatest warming, over many eastern U.S. forests.
- Research results suggest that increasing atmospheric CO_2 levels could stimulate the growth of rangeland plant species because of the direct CO_2 fertilization effect and indirectly by reducing water stress by virtue of increased water-use efficiency of plants at elevated CO_2 levels. In rangelands where undesirable species such as mesquite occur or are introduced, however, elevated CO_2 also could cause a deleterious effect on rangeland plant species composition by increasing the growth and establishment of such species.

Table 7
Understanding Changes in Ecosystems

FY 2001 Budget by Agency
(discretionary budget authority in $millions)

USDA		29.2
DOE		10.8
DOI/USGS		13.9
EPA		3.0
NASA	Scientific Research	32.0
NASA	Space-Based Observations	101.9
NSF		29.0
Smithsonian Institution		3.8
TOTAL		**223.6**

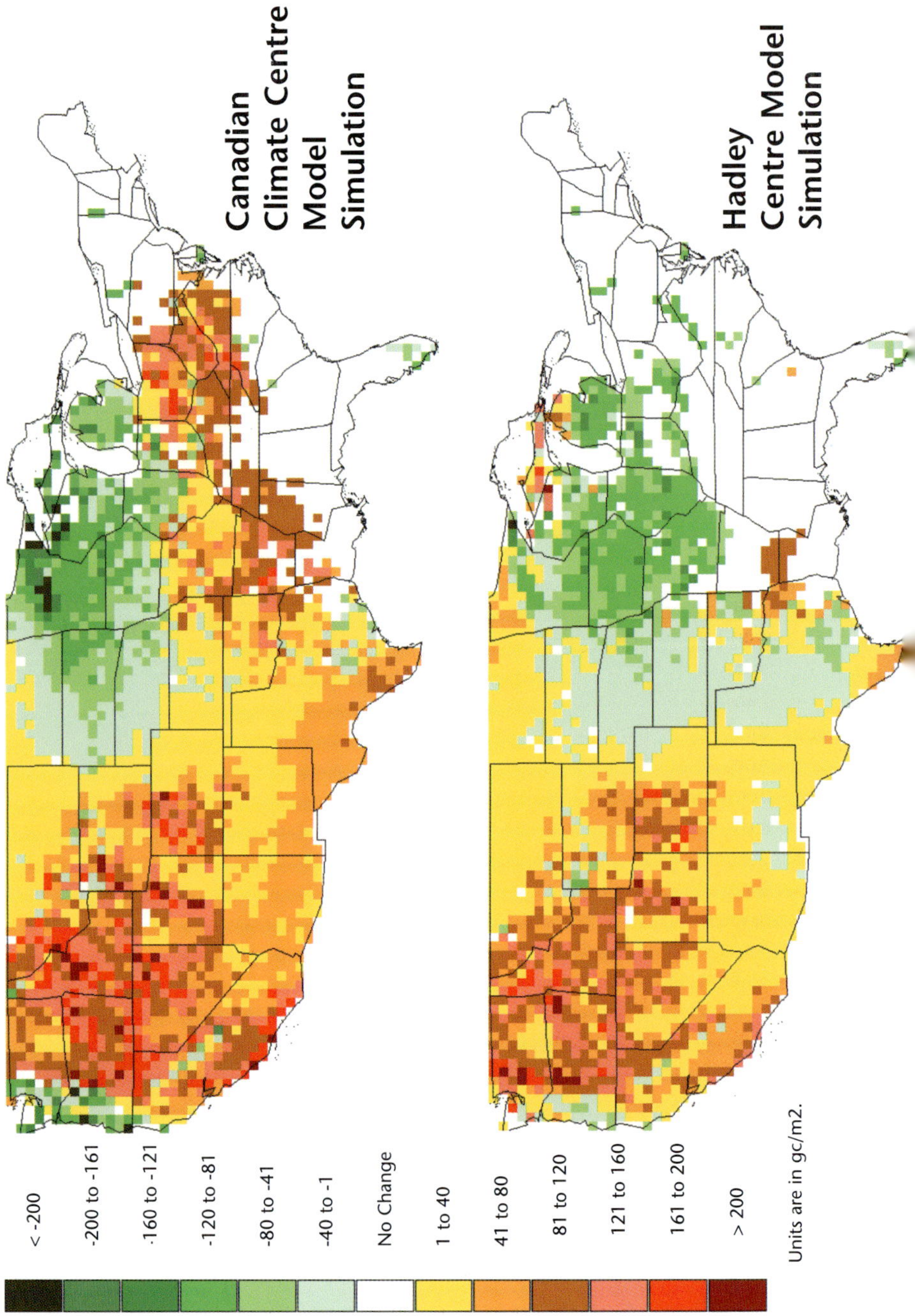

Figure 5. Potential Change in Wildfire Under Climate Change

(See pages 41-42 for additional information)

FY 2001 Plans

The USGCRP will increase its emphasis on research to explain the relationships among land cover, land use, biodiversity, climate, and weather, and investigate ecosystem resilience to single and multiple stresses. Existing and new observational capabilities will be used to support increased understanding of ecosystems functioning and processes. Goals include:

- Producing quantitative estimates of the rates of land-use change and its effects on atmospheric chemistry and biodiversity in tropical areas such as Central Africa, Southeast Asia, and previously unmeasured areas of South America (non-Brazilian Amazon).
- Developing and publishing (electronically via the USGCRP Web site) of criteria and indicators of ecosystem sensitivity and sustainability to link landscape-scale changes and disturbance regimes with ecosystem carrying capacity.
- Producing an assessment of the state of understanding of advanced ecological and biogeochemical cycling models that blend processes occurring across time spans of days to centuries and seek to portray realistically the effects of disturbance and land-use history.
- Successful collection, processing, and analysis of global measurements of land and ocean surface temperatures, ocean color, ocean chlorophyll fluorescence, and terrestrial vegetation from the MODIS instrument on NASA's EOS-Terra satellite. Terra-MODIS, together with the SeaWiFS satellite, will provide significantly improved coverage of key elements of land and ocean ecosystems and thereby improve statistics on the variability of these systems.
- Creating and regularly updating the first high-resolution satellite data sets of tropical biodiversity hotspots using data from the Landsat 7 satellite, which includes a new 15 m resolution panachromatic band for the study of ecosystem disturbances. The USGCRP will continue the acquisition of Landsat-7 measurements and improve the land use/land cover data series provided by other Landsat missions for 26 years.

Understanding the Human Dimensions of Global Change

The USGCRP budget includes $95 million in FY 2001 for the study of human dimensions of global change (see Table 8). About half of this amount is devoted to improving our understanding of the human health effects of stratospheric ozone depletion. Scientific uncertainties about the role of human socioeconomic and institutional factors in global change are as significant as uncertainties about physical, chemical, and biological aspects of the Earth system. Improving our scientific understanding of how humans cause changes in the Earth system—and how society, in turn, is affected by the interactions between natural and social processes—is an important priority for the USGCRP. Key questions include the following: What are the major human drivers of changes and how do they vary temporally, spatially, and across economic sectors and social groups? What are the human consequences of global environmental change? How might global change affect key life-support systems (water, health, agriculture), economies, and political systems?

Although systematic research on these questions is still at a relatively early stage of development, USGCRP human dimensions research efforts have resulted in insights that enhance our understanding of and ability to respond to the multiple stresses within our environment. Results from these efforts have begun to influence studies across the physical and biological sciences; as a result, such information and knowledge are becoming increasingly accessible to individuals and institutions who are responsible for action in areas that are sensitive to climate and environmental change.

The USGCRP will increase its research on the processes through which individuals and organizations perceive, identify, explore alternatives, and respond to a range of environmental hazards and risks—including sudden disasters such as earthquakes, severe storms, and floods and longer-term degradation of environmental quality. The USGCRP will communicate its research results to relevant decisionmakers in disaster management, engineering, and economic development.

Recent Accomplishments

- NOAA, NSF, EPA, NASA, and the Electric Power Research Institute initiated and produced the first joint research announcement on Climate Variability and Health. This announcement was released in November 1999, and proposals are now undergoing evaluation. NOAA made significant progress in overcoming legal hurdles to allow a high degree of interagency cooperation and private sector involvement.
- Interdisciplinary investigations of human responses to seasonal and yearly swings in climate are highlighting the effects of market forces, access to resources, institutional flexibility, transboundary context, and the role of indigenous knowledge on the likelihood that individuals and institutions will use improved scientific information.
- Integrated assessment studies are increasing our understanding of the relationship between climate change and human responses. Generally, scientists have found that people fear projected future changes in their environments—but that small events and slowly-changing environmental conditions rarely generate levels of concern that lead people to adopt new adaptation or mitigation strategies.
- EPA's Global Change Research Program issued its first Request for Assistance (RFA) related to human dimensions issues. This RFA, titled "Assessing the Consequences of Interactions between Human Activities and a Changing Climate," represents the first step in a multi-year plan to foster the development of models that capture the interaction of human behavior with natural (nonhuman) responses to climate change. Proposals are now undergoing peer review.
- The U.S. Department of Transportation recently established a virtual Center for Climate Change and Environmental Forecasting to identify effective ways to reduce transportation sector emissions and help prepare the Nation for the impacts of climate change. As part of its research efforts, the Center plans to investigate how climate change might affect transportation infrastructure.

Table 8
Understanding the Human Dimensions of Global Change

FY 2001 Budget by Agency
(discretionary budget authority in $millions)

DOC/NOAA	Scientific Research	5.2
DOE		7.8
EPA		19.7
HHS/NIH		48.0
NSF		14.0
Smithsonian Institution		0.6
TOTAL		**95.3**

FY 2001 Plans

The USGCRP will continue to support fundamental research on and assessments of the effects of human activities on the environment and of the potential human consequences of global change. Goals include:

- Initiating new assessments of the effects of global change on weather-related morbidity, aquatic ecosystems, and the consequences of global change for tropospheric ozone.
- Completing and publishing eight regional assessments of the potential consequences of climate variability and change in the United States. These assessment activities were sponsored by the USGCRP as part of its National Assessment process. (Eight regional assessments and five sector assessments are expected to be completed by the end of FY 2000 and published as reports or in journals.)
- Examining the social and economic cost/benefit implications of long-term future greenhouse gas emission scenarios and consequent environmental change by conducting an intercomparison of more than 10 integrated assessment models.
- Assessing the health effects of combined exposures to climatic and environmental factors, with a focus on improving our understanding of the combined effects of higher temperatures and air pollutant concentrations.
- Continuing efforts to determine the effects of depletion of stratospheric ozone and providing this information to policymakers. This effort will include focused research on the mechanisms of skin cancers induced by UV-B radiation, the effects of UV-B radiation on the systemic and cutaneous immune systems, and the relationship of these effects to skin cancer.

Paleoclimate: The History of the Earth System

The USGCRP budget includes $27 million in FY 2001 for the study of the Earth's amazingly complex climate and environmental history (see Table 9). This element of the USGCRP focuses on providing a quantitative understanding of how the environment has changed in the past and defining the envelope of natural environmental variability within which the effects of human activities on the planet's biosphere, geosphere, and atmosphere can be assessed.

Paleoenvironmental records are derived from a wide variety of natural archives—such as: lake and ocean sediments, tree rings, wind-blown deposits, coral, and ice cores—as well as historical documents. Chemical, isotopic, and ecological analyses of these records have demonstrated that the natural climate system has varied locally and globally over a far greater range than can be inferred from relatively short-term instrumental records. In most locations, instrumental records might provide 100 years of climate data, whereas an ice core might provide an annual climate record of 10,000 to 30,000 years (more than 400,000 years in Antarctica).

Understanding the natural environmental changes of our planet on long timescales (years to millennia) provides the context for understanding today's climate dynamics and elucidating the effects of natural versus anthropogenic influences. Reconstructing past climate records offers an enhanced understanding of mechanisms that control the Earth's climate system and, together with insight from numerical modeling exercises, provides a foundation for anticipating how the planet might respond to future environmental perturbations.

Recent Accomplishments

- Recent progress in synthesizing various proxy records of past climates enables placement of 20th century climate warming within a longer-term perspective. Recent results indicate that much of the variability during the past 1,000 years prior to the rise in emissions of greenhouse gases from human activities (beginning in about 1850) can be attributed to pulses of volcanism or changes in the output of the Sun's energy. Neither of these mechanisms—or natural climate variability in the ocean-atmosphere system—can explain the late 20th century rise in the globally averaged surface temperature. Greenhouse gases appear to emerge as the dominant forcing during the 20th century. According to proxy temperature records, the 1990s appear to have been the warmest decade (and 1998 the warmest year) in the past 1,000 years.
- A partnership between two sets of researchers has resulted in the acquisition of ice core and meteorological data from the Sajama ice cap in Bolivia. This research has provided a record of glacial versus interglacial tropical climate dynamics over the past 25,000 years and is contributing to a better understanding of past and present tropical Pacific climate variability.

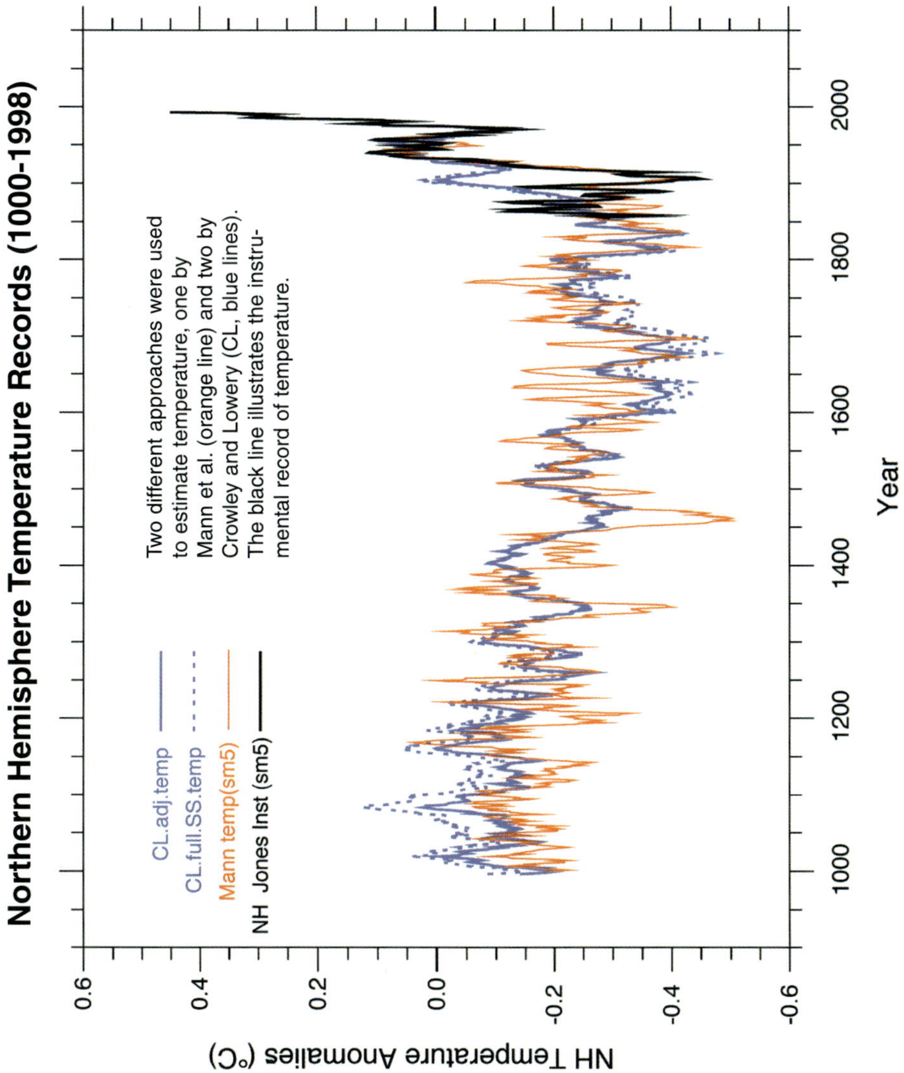

Figure 6. Northern Hemisphere Temperature Records for the Past 1,000 Years

(See page 41 for additional information)

Table 9
Paleoclimate: The History of the Earth System

FY 2001 Budget by Agency
(discretionary budget authority in $millions)

DOC/NOAA Scientific Research	2.3
DOI/USGS	4.0
NSF	19.5
Smithsonian Institution	1.6
TOTAL	**27.4**

FY 2001 Plans

The USGCRP will focus on acquisition of new paleoenvironmental data, enhancing paleoclimate modeling, and improving access to both data and models. Goals include:

- Constructing new long-term climate and environmental records for parts of North and South America by successfully obtaining and analyzing eight new long, continuous sediment cores—four from the Great Salt Lake in the United States and four from Lake Titicaca in Bolivia. This will be accomplished by using a newly acquired drilling system for obtaining sediment cores from beneath large lakes. These sedi ments are the primary source of long-term paleoenvironmental records from conti nental areas. Acquisition and study of such sediments has been the object of years of planning and coordination work.
- Increasing overall paleoclimate modeling activities by 30-50 percent by improving computing capabilities and providing model access to additional users. Data-model c omparisons will be an important part of these activities. This paleo-modeling effort will be complementary to, and compatible with, the extensive climate simulation efforts for modern climate.
- Doubling the overall volume of publicly accessible paleoclimate data holdings, expanding the utility of these data to facilitate their application by a wider range of users, and expanding data access to 10 percent more users.

INTERNATIONAL CONNECTIONS

The USGCRP addresses an important set of national objectives; it also participates in a series of international efforts directed toward improving understanding of change on global and regional scales. The resulting databases, when aggregated, provide essential inputs to the increasingly complex coupled models that enable scientists to improve analysis and prediction of global change. Regular interaction between the USCGRP and these international scientific efforts also results in substantial feedback to our assessment of the impacts of global change and their consequences for the North American continent. Some examples of USGCRP international connectivity of special relevance for scientific areas that the USGCRP expects to emphasize in the coming year are described below.

International Global Change Research Programs

At the global level the **International Geosphere-Biosphere Programme (IGBP)** has as its primary goal describing and understanding the interactive physical, chemical, and biological processes that regulate the Earth system. The IGBP will convene a major International Open Science Meeting in July 2001 to review the progress of IGBP research; synthesize results achieved; and set directions for future IGBP research in areas such as global atmospheric chemistry, global change and terrestrial ecosystems, and land-use and land-cover change.

The **World Climate Research Programme (WCRP)** is directed at improving scientists' understanding of the Earth's climate system and climate processes and, through such studies, determining the extent to which climate can be predicted and the extent to which humankind influences climate. The WCRP's Global Energy and Water Cycle Experiment (GEWEX) will be of increasing importance to U.S. scientists as the USGCRP increasingly emphasizes research on changes in the global water cycle as a primary determinant of the Earth's climate.

The **International Human Dimensions Programme (IHDP)** coordinates multidisciplinary international collaborative scientific programs that examine the causes and environmental consequences of people's individual and collective actions as well as the potential efficacy and impact of various general adaptation and mitigation strategies. The IHDP also works to identify emerging research opportunities and provides synthesis reports and policy-oriented summaries to contribute a scientific basis for individual and collective decisionmaking.

Regional Activities

At the regional level, the **International Research Institute for Climate Prediction (IRI)** prepares, issues, and distributes interannual climate forecasts, such as forecasts of El Niño events, derived from global and regional ocean-atmosphere coupled models. The primary audience is countries that are especially vulnerable to global change and the major interannual events associated with such change. IRI forecasts are of special value in sectors such as agriculture, health, and water resources.

The **Inter-American Institute for Global Change Research (IAI)** is an intergovernmental treaty organization established in 1994 to promote collaborative research to understand and predict the integrated impact of present and future global changes. The IAI has established a network of more than 200 collaborating institutions throughout the

Americas, involving nearly 1,000 scientists. The results of IAI projects to date have demonstrated that high-quality, peer-reviewed science, along with capacity building, can produce broad benefits.

The **Asia-Pacific Network for Global Change Research (APN)** made notable progress in 1999 and was able to increase substantially its support for climate research and studies of the human dimensions of global change. U.S. scientists are substantially involved in APN research programs.

Bilateral Cooperation

U.S. scientists involved in global change research also work closely with their counterparts in many other countries on a bilateral basis. The USCGRP strongly encourages such direct scientific cooperation. Of special note is the growing cooperation between U.S. and Japanese scientists in several areas of mutual interest.

In 1999, the Seventh **U.S.-Japan Workshop on Global Change Research** focused on Precipitation Systems and Their Variability in the Asia-Pacific Region. This workshop resulted in several recommendations for follow-up, some of which are being pursued actively this year. The eighth in this series of workshops, which will address Global Change and Environment, will be convened in November 2000 at NIH. Special areas of emphasis for this workshop will be climate change and air pollution impacts on human health, and health impacts of stratospheric ozone depletion and greater exposure to harmful solar radiation.

International Interactions in Support of Observations and Monitoring

USGCRP scientists are actively engaged in several efforts to expand and improve U.S. capabilities to observe and monitor the Earth system in support of global change research. At the international level, the **Committee on Earth Observational Satellites (CEOS)** provides a forum through which the United States and other countries that are conducting Earth remote-sensing programs coordinate their activities. This forum is of special importance now because of the successful launch of a series of U.S. Earth remote-sensing satellites in 1999. Cooperative arrangements with the European Space Agency (ESA) and the National Space Development Agency of Japan (NASDA) also are valuable in this area.

The **Integrated Global Observing Strategy (IGOS)** Partnership brings together CEOS; the International Group of Funding Agencies for Global Change Research (IGFA); the IGBP and WCRP; and United Nations agencies involved in in situ ocean, atmosphere, and terrestrial observing systems to promote effective interaction and complementarity between remote sensing and in situ observing systems for global change research and prediction.

International Scientific Assessment

The Intergovernmental Panel on Climate Change (IPCC) was established in 1988 by the World Meteorological Organization (WMO) and the United Nations Environment Programme (UNEP) to assess available scientific, technical, and socioeconomic information in the field of climate change. A substantial number of USGCRP-supported researchers are contributing to the IPCC's Third Assessment Report on the science of climate change; climate change impacts, adaptation, and vulnerability; and mitigation of climate change.

RECENT ASSESSMENTS AND REPORTS ON GLOBAL CHANGE

Intergovernmental Panel on Climate Change. *Climate Change 1995: The Science of Climate Change*. Contribution of Working Group I to the Second Assessment Report of the IPCC. Cambridge, U.K.: Cambridge University Press, 1996.

Intergovernmental Panel on Climate Change. *Climate Change 1995: Impacts, Adaptations and Mitigation of Climate Change: Scientific-Technical Analyses*. Contribution of Working Group II to the Second Assessment Report of the IPCC. Cambridge, U.K.: Cambridge University Press, 1996.

Intergovernmental Panel on Climate Change. *Climate Change 1995: Economic and Social Dimensions of Climate Change*. Contribution of Working Group III to the Second Assessment Report of the IPCC. Cambridge, U.K.: Cambridge University Press, 1996.

Intergovernmental Panel on Climate Change. *The Regional Impacts of Climate Change: An Assessment of Vulnerability*. Special Report of the IPCC. Cambridge, U.K.: Cambridge University Press, 1997.

Intergovernmental Panel on Climate Change. *Aviation and the Global Atmosphere*. Special Report of the IPCC. Cambridge, U.K.: Cambridge University Press, 1999.

Intergovernmental Panel on Climate Change. *Land Use, Land-Use Change, and Forestry*. Special Report of the IPCC. Cambridge, U.K.: Cambridge University Press, 2000

Intergovernmental Panel on Climate Change. *Emissions Scenarios*. Special Report of the IPCC. Cambridge, U.K.: Cambridge University Press, 2000.

Intergovernmental Panel on Climate Change. *Methodological and Technological Issues in Technology Transfer*. Special Report of the IPCC. Cambridge, U.K.: Cambridge University Press, 2000.

National Research Council. *Global Environmental Change: Research Pathways for the Next Decade*. Washington, D.C.: National Academy Press, 1999.

National Research Council. *Reconciling Observations of Global Temperature Change*. Washington, D.C.: National Academy Press, 2000.

United Nations Environment Programme (UNEP). *Synthesis of the Reports of the Scientific, Environmental Effects, and Technology and Economic Assessment Panels of the Montreal Protocol: A Decade of Assessments for Decision Makers Regarding the Protection of the Ozone Layer: 1988-1999*. Nairobi, Kenya: UNEP, 1999. (Copies of this report are available from the Web site of the UNEP Ozone Secretariat at: http://www.unep.org/ozone or http://www.unep.ch/ozone.)

United Nations Environment Programme (UNEP). *Environmental Effects of Ozone Depletion: 1998 Assessment*. Nairobi, Kenya: UNEP, 1998. (Copies of this report are available at the following Web site: http://sedac.ciesin.org/ozone/docs/UNEP98/UNEP98.html.)

U.S. Global Change Research Program. *A U.S. Carbon Cycle Science Plan*. Report of the Carbon and Climate Working Group, 1999.

U.S. Global Change Research Program. *1999 Newly Available Agency Data Sets That are Significantly Global Change Related*. Report of the USGCRP Data Management Working Group. (This report is available on the Global Change Data and Information System Web site at: http://globalchange.gov/data/datasets-1999.html.)

World Meteorological Organization, United Nations Environment Programme, NOAA, NASA, and European Commission. *Scientific Assessment of Ozone Depletion: 1998*. (This report may be ordered on the UNEP Web site at: http://www.unep.org/ozone or http://www.unep.ch/ozone.)

FIGURE CAPTIONS

Figure 1: Asymmetric Trends in Arctic and Antarctic Sea-Ice Extent

General circulation model (GCM) experiments that simulate future climate conditions assuming a gradual increase in atmospheric CO_2 show various hemispheric asymmetries in global sea ice extent. Some suggest that Arctic sea ice will decrease significantly, whereas Antarctic sea ice will decrease substantially less or even increase. Scientists at the NASA Goddard Space Flight Center have found observational evidence that there is a current asymmetry in global sea ice changes. In particular, they found that the areal extent of sea ice decreased by 2.9 +/- 0.4% per decade in the Arctic and increased by 1.3 +/- 0.2% per decade in the Antarctic from November 1978 through December 1996. These results are based on measurements from passive microwave instruments on the Nimbus 7 and three defense Meteorological Satellite Program spacecraft. The assymetry in sea ice trends is intriguing and indicates that climate in the southern polar region may have very different dynamics from the climate in the northern polar region.

Source: NASA Goddard Space Flight Center. D.J. Cavalieri, P. Gloersen, C.L. Parkinson, J.C. Comiso, and H.J. Zwally, Laboratory for Hydrospheric Processes/Code 971.

Figure 2a: Polar ozone loss is now evident in both hemispheres, raising concern that climate change could delay the recovery of stratospheric ozone

Figure 2a is a graph of total column ozone (in Dobson units) as a function of time since 1970, as obtained from several NASA and NOAA satellites. The points plotted represent a monthly average of total column ozone for March in the Northern Hemisphere (NH) and October in the Southern Hemisphere (SH). The plotted points also represent an average of data obtained from 63 to 90 degrees latitude (defined as the polar region) in each hemisphere. In the SH polar springtime (October), stratospheric ozone began a steep decline in 1980 that has continued almost smoothly, leading to steady, severe loss over the last decade. In the NH springtime, the situation is different. Stratospheric ozone levels were fairly constant during the 1980s, with a significant decline not beginning until 1990. The NH decline also exhibits greater interannual variability. Both SH and NH responses are due to the increase in human-supplied chlorine to the stratosphere from ozone-destroying industrial chemicals now banned by the Montreal Protocol and its amendments and adjustments. The years with severe NH polar ozone loss are strongly correlated with NH winters of colder than average stratospheric temperatures, with the cold persisting until the sunlight returns in the springtime. These are the usual conditions in SH, but until recently have been uncommon the NH. An emerging explanatory paradigm is that greenhouse gas forcing has been cooling the NH polar lower stratosphere through dynamical and radiative mechanisms. This interaction of stratospheric chemistry and climate change could delay the expected recovery of stratospheric ozone by a number of decades.

Source: NASA

Figure 2b and 2c: Stratospheric ozone is depleted, as the level of active chlorine (ClO) rises, along the flight track of NASA's ER-2 high-altitude airborne laboratory

Figure 2b and 2c are two plots of measurements made by the instruments on board NASA's ER-2 high-altitude aircraft. These observations were made during field campaigns designed to explore in detail the health of the stratospheric ozone layer in the Antarctic, the Airborne Antarctic Ozone Experiment (AAOE, 1987), and in the Arctic, the SAGE III Ozone Loss and Validation Experiment (SOLVE, 1999-2000).

Figure 2b shows ozone and chlorine monoxide concentrations in the lower stratosphere (number of ozone (O_3) and chlorine monoxide (ClO) molecules per billion molecules of air (in blue and red, respectively) plotted against Greenwhich mean time during the AAOE ER-2 flight of September 16, 1987. Chlorine monoxide is the catalytically active radical responsible for severe polar ozone loss in the polar springtime. As the aircraft flight track crosses into the Antarctic polar vortex, the concentration of ClO goes up by a factor of 10 while ozone concentration decreases by more the 60% during this same period of the flight. The anti-correlation between ozone and ClO concentrations, exhibited in both the large-scale changes and the smaller fluctuations, is a strong indication that anomalous chlorine chemistry is responsible for the ozone depletion.

In Figure 2c, a similar plot has been constructed for an ER-2 flight in the Arctic during the SOLVE campaign conducted on March 11, 2000. Again we see the anti-correlation between ozone and ClO, this time as the plane flies into the Arctic polar vortex. With advances in measurement capability, we can now see additional important atmospheric species, chlorine nitrate ($ClONO_2$), the important reservoir molecule that stores the human-supplied chlorine in a form not directly destructive to ozone, and chlorine peroxide (Cl_2O_2), the dimer of ClO that is the source of the destructive ClO when the sunlight returns to the polar regions in the spring. These new observations, a benefit of technological advances since AAOE, help form a more complete picture of the complex chemistry of polar ozone loss, and have confirmed that this chemical ozone loss occurs in the Arctic.

Atmospheric chlorine, a result of human emissions, has been available in the atmosphere for many years, and conversion from the innocuous reservoir forms (such as chlorine nitrate) to the destructive radical forms (ClO) in the early polar spring has been a regular feature of springtime Antarctic chemistry. However, the extremely cold stratospheric temperatures, a necessary ingredient for this conversion, have also started in recent years to become a more regular feature in the Arctic springtime. This development, a possible consequence of increasing greenhouse gases, has the potential for significant impacts on the larger population of the Northern Hemisphere.

Source: NASA

Figure 3: TRMM Measures Sea Surface Temperature Through Clouds

This image was acquired over the tropical Atlantic Ocean and U.S. East Coast regions from Aug. 22 - Sept. 23, 1998. Cloud data were collected by the Geostationary Operational Environmental Satellite (GOES). Sea Surface Temperature (SST) data were collected aboard the NASA/NASDA Tropical Rainfall Measuring Mission (TRMM) satellite by the TRMM Microwave Imager (TMI). TMI is the first satellite microwave sensor capable of accurately measuring sea surface temperature through clouds, with results shown in this scene.

For years scientists have known there is a strong correlation between sea surface temperature and the intensity of hurricanes. But one of the major stumbling blocks for forecasters has been the precise measurement of those temperatures when a storm begins to form. In this scene, clouds have been made translucent to allow an unobstructed view of the surface. Notice Hurricane Bonnie approaching the Carolina Coast (upper left) and Hurricane Danielle following roughly in its path (lower right). The ocean surface has been falsely colored to show a map of water temperature—dark blues are around 75°F, light blues are about 80°F, greens are about 85°F, and yellows are roughly 90°F.

A hurricane gathers energy from warm waters found at tropical latitudes. In this image we see Hurricane Bonnie cross the Atlantic, leaving a cooler trail of water in its wake. As Hurricane Danielle followed in Bonnie's path, the wind speed of the second storm dropped markedly, as available energy to fuel the storm dropped off. But when Danielle left Bonnie's wake, wind speeds increased due to temperature increases in surface water around the storm.

As a hurricane churns up the ocean, it's central vortex draws surface heat and water into the storm. That suction at the surface causes an upwelling of deep water. At depth, tropical ocean waters are significantly colder than water found near the surface. As they're pulled up to meet the storm, those colder waters essentially leave a footprint in the storm's wake which might last as long as two weeks. Forecasters can quantify the difference in surface temperatures between this footprint and the surrounding temperatures and use that information to better predict storm intensity. If another storm intersects with this cold water trail, it is likely to lose significant strength due to the fact that the colder water does not contain as much potential energy as warm water.

Source: TRMM Project, Remote Sensing Systems, and Scientific Visualization Studio, NASA Goddard Space Flight Center
Image may be viewed at:
http://earthobservatory.nasa.gov/Newsroom/NewImages/images.php3?img_id=2918

Figure 4: Partitioning of Fossil-Fuel-Derived Carbon

Annual sources and sinks of atmospheric CO_2 as determined from atmospheric measurements of CO_2 and its isotopic ratio $^{13}C/^{12}C$. Red: fossil fuel combustion; blue: net uptake by the oceans (sometimes a source); green: net uptake by terrestrial ecosystems (a source due to tropical deforestation is included in these numbers); gray: the measured atmospheric increase, which is the sum of fossil fuel combustion and the exchanges with the oceans and terrestrial biosphere. The atmosphere increases every year.

Sources: Francey et al., Nature 373, 326-330 (1995); Tans et al. (private communication), updated from Ciais et al., Science 269, 1098-1102, (1995); see also www.cmdl.noaa.gov (Carbon Cycle Greenhouse Gases).

Figure 5: Potential Change in Wildfire Over the Conterminous U.S. Under Climate Change

The potential change in biomass consumed by fire is shown for two future climate scenarios as simulated by the MC1 dynamic general vegetation model. The future climate scenarios are from the Hadley Centre (HADCM2SUL) and the Canadian Climate Centre (CGCM1). Shown is the difference in the average annual carbon consumed by wildfire over the past 100 years compared to that simulated over the next 100 years under each scenario. The simulations indicate that climate change could lead to increased wildfire over much of the western U.S. The green colors show a reduction in biomass consumed by fire (kg dry matter per m^2/yr, while the yellows to reds indicate an increase in biomass consumed. Both the Hadley and Canadian future climate scenarios produce increases in biomass consumed in the West, particularly the Southwest and interior dry forests of the Great Basin. Increased precipitation under both scenarios throughout much of the West produces more vegetation, which adds fuel to relatively dry ecosystems. Occasional dry years, thus produce more and larger fires in the drier Southwest and Great Basin ecosystems. Increased drought stress in the Southeast forested region under the Canadian scenario also results in more and larger fires in that region. These fire impacts could begin occurring within the next few decades.

MC1 (MAPSS-CENTURY, version 1, Daly et al. 2000) is a dynamic general vegetation model, combining shifting vegetation distribution from the MAPSS biogeography model (Mapped Atmosphere-Plant-Soil System, Neilson 1995) with carbon and nutrient dynamics from the CENTURY biogeochemistry model (Parton et al. 1987). MC1 also contains a state-of- the-art, process-based fire model (Lenihan et al. 1998), which calculates fuel loads, moisture levels, and wildfire as a function of climate and vegetation characteristics. These results are for "potential dynamic vegetation" only and do not include land-management activities, such as fire suppression, forest harvest, or conversion to agriculture.

Source: Ronald P. Neilson (USDA Forest Service); Dominique Bachelet, James M. Lenihan and Raymond J. Drapek (Oregon State University).

Bachelet, D., Neilson, R. P., Lenihan, J. M., and Drapek, R. J. (in review), "Climate Change Effects on Vegetation Distribution and Carbon Budget in the U.S." Submitted to: *Ecosystems*. (This journal paper is one of several submitted as a group to Ecosystems as background documents to the USGCRP National Climate Assessment.)

References:

Daly C., Bachelet D., Lenihan J.M., Neilson R.P., Parton W., and Ojima D. 2000. Dynamic simulation of tree-grass interactions for global change studies. *Ecological Applications* 10(2) 449-469.

Lenihan J.M., Daly C., Bachelet D., and Neilson, R.P.. 1998. Simulating broad-scale fire severity in a dynamic global vegetation model. *Northwest Science* 72:91-103.
Neilson R.P. 1995. A model for predicting continental scale vegetation distribution and water bal-

ance. *Ecological Applications* 5(2):362-385.

Parton W.J., Schimel D.S., Cole C.V., and Ojima D. 1987. Analysis of factors controlling soil organic levels of grasslands in the Great Plains. *Soil Science Society of America* 51:1173-1179.

Figure 6: Northern Hemisphere Temperature Records for the Past 1,000 Years

Comparison of different Northern Hemisphere temperature reconstructions for the last 1000 years. Two different approaches were used to estimate temperature, one by Mann et al. (orange line) and two by Crowley and Lowery (CL, blue lines). The solid blue line is the best CL estimate; the dashed line represents an estimate incorporating lower resolution data that are not as reliable. The black line illustrates the instrumental record of temperature. Regardless of which approach was adopted, all reconstructions (plus two others for midlatitude summers) agree that late 20th century temperatures are the highest in at least the last 1000 years.

Source: Figure from Thomas J. Crowley and Thomas S. Lowery, "How Warm Was the Medieval Warm Period?" *Ambio*, Vol. 29, No. 1, Feb. 2000, pp. 51-54 (Royal Swedish Academy of Sciences). Figure also uses data from: (1) Mann, M.E., Bradley, R.S., and Hughes, M.K., 1999, "Northern Hemisphere Temperatures During the Past Millennium: Inferences, Uncertainties, and Limitations," *Geophys. Res. Lett.*, 26, 759-762; and (2) Jones, P.D., New, M., Parker, D.E., Martin, S., and Rigor, I.G., 1999, "Surface Air Temperature and its Changes Over the Past 150 Years," *Rev. Geophys.*, 37, 173-199.

APPENDIX: THE FY 1999-2001 USGCRP BUDGET BY AGENCY AND PROGAM

The first table in this Appendix presents the FY 2000-FY 2001 USGCRP budget by Program Element, showing each agency's budgetary contribution to each element. The budget pages for individual participating departments and agencies that follow include a listing of programs designated for inclusion in the USGCRP, as well as a general description of each agency's "Areas of Global Change Research." For each agency, a "FY 2001 Program Highlights" section outlines briefly some of the key USGCRP-related activities proposed for the coming year. In addition, the agencies conduct a broad range of "Related Research," as indicated, funding for which is not included as part of the USGCRP budget because the research is conducted primarily for other purposes.

The resources allocated to specific programs within agencies as reflected in these tables for FY 2000 appropriated funds and the FY 2001 budget request are estimates only, and are subject to change based on decisions on scientific and programmatic priorities among USGCRP agencies and their advisory bodies and on the input of the national and international scientific communities.

Each agency budget also includes a "Mapping of Budget Request to Appropriations Legislation." The entry for each department or agency points to the location (or locations) in the various Appropriations bills (and, in some cases, Appropriations Committee reports) of funding for USGCRP activities. Note that it is common for global change research to be funded within Appropriations accounts that also include funding for other activities, so that Appropriations bills and committee reports do not necessarily designate funding specifically for global change research. Thus, the actual funding level for global change research activities must be determined, in part, by decisions within agencies about how to allocate appropriated funds. It should also be noted that global change research activities are funded by seven separate Appropriations bills. Thus, the relationship between the USGCRP budget and the Appropriations process is complex and not easily summarized.

U.S. Global Change Research Program

FY 2000 – FY 2001 Budget by Program Element by Agency
(Discretionary budget authority in $millions)

AGENCY	DOC/NOAA Science	DOC/NOAA Obs.	DOE	DOI/ USGS	EPA	HHS/ NIH	NASA Obs.	NASA Science	NSF	SI	USDA	Total
Program Element												
Understanding the Climate System												
FY00	41.1	0.0	63.2	0.0	0.0	0.0	251.4	58.8	83.7	0.4	0.0	498.6
FY01	41.1	19.8	69.5	0.0	0.0	0.0	208.8	62.3	83.5	0.4	0.0	485.4
Understanding the Composition and Chemistry of the Atmosphere												
FY00	8.8	0.0	14.6	0.0	0.0	0.0	273.7	56.4	18.7	0.3	15.4	387.9
FY01	8.8	1.0	12.6	0.0	0.0	0.0	252.7	53.7	18.7	0.3	17.6	365.4
Global Water Cycle												
FY00	5.0	0.0	4.0	0.0	0.0	0.0	205.4	49.4	9.7	0.0	0.3	273.8
FY01	5.0	1.9	3.0	0.0	0.0	0.0	228.7	59.5	9.7	0.0	0.3	308.1
Global Carbon Cycle												
FY00	4.6	0.0	13.3	3.3	0.0	0.0	113.6	40.6	13.1	0.3	15.4	204.2
FY01	4.6	5.3	15.6	3.5	0.0	0.0	104.6	45.0	13.1	0.3	37.2	229.2
Understanding Changes in Ecosystems												
FY00	0.0	0.0	10.8	13.4	3.0	0.0	93.0	30.6	29.0	3.8	22.0	205.6
FY01	0.0	0.0	10.8	13.9	3.0	0.0	101.9	32.0	29.0	3.8	29.2	223.6
Understanding the Human Dimensions of Global Change												
FY00	5.2	0.0	8.2	0.0	17.6	46.0	0.0	0.0	14.0	0.6	0.0	91.6
FY01	5.2	0.0	7.8	0.0	19.7	48.0	0.0	0.0	14.0	0.6	0.0	95.3
Paleoclimate: The History of the Earth System												
FY00	2.3	0.0	0.0	6.4	0.0	0.0	0.0	0.0	19.3	1.6	0.0	29.6
FY01	2.3	0.0	0.0	4.0	0.0	0.0	0.0	0.0	19.5	1.6	0.0	27.4
Agency Totals												
FY00	**67.0**	**0.0**	**114.1***	**23.0***	**20.6**	**46.0**	**937.1**	**235.8**	**187.5**	**7.0**	**53.1**	**1,691.4***
FY01	**67.0**	**28.0**	**119.3***	**21.4***	**22.7**	**48.0**	**896.7**	**252.5**	**187.5**	**7.0**	**84.6**	**1,734.7***

* Excludes $4.0 million for DOI/USGS data management and $3.1 million for the DOE SBIR/STTR program in FY 2000 and FY 2001.

Department of Agriculture

USDA	Program Title	FY99	FY00 Estimate	FY01 Request
ARS	Agriculture and Rangeland Global Change	11.4	11.4	14.4
ARS	Methyl Bromide Research	14.4	14.4	16.0
ARS	Carbon Cycle Research	0.0	0.9	5.0
CSREES	Improved Response Models	6.4	6.4	11.9
CSREES	UV-B Monitoring Network	1.0	1.0	1.6
ERS	Economic Incentive Carbon Sequestration	0.0	0.0	0.7
ERS	Economics of Global Change and Agriculture	0.8	0.8	0.8
FS	Forest Global Change Research	16.9	16.9	16.9
FS	Carbon Cycle Science	0.0	0.0	3.0
NRCS	Soil Carbon Studies	1.2	1.2	1.5
NRCS	Soil Surveys and Inventories	0.0	0.0	12.0
ARS	Regional, Sectoral, and National Assessments	0.0	0.0	0.8
	USDA Total	**52.1**	**53.0**	
	President's Request			**84.6**

ARS	Agricultural Research Service
CSREES	Cooperative State Research, Education, and Extension Service
ERS	Economic Research Service
FS	Forest Service
NRCS	Natural Resources Conservation Service

Areas of Global Change Research. USDA-sponsored research focuses on understanding terrestrial systems and the effects of global change (including water balance, atmospheric deposition, vegetative quality, and UV-B radiation) on food, fiber, and forestry production in agricultural, forest, and range ecosystems and examines how agricultural and forestry activities can contribute to a reduction in greenhouse gases.

FY 2001 Program Highlights. In FY 2001, USDA will increase its carbon cycle research program. As part of the interagency Carbon Cycle Science Program, USDA will collaborate with other Federal agencies to conduct research to better understand how agricultural practices affect the net carbon balance and develop methods which will assist farmers, ranchers, and forest landowners to increase carbon sequestration. Special emphasis will be given to measurement of the effects of management and conservation practices on carbon storage in cropland and grazing lands. Basic research will define the mechanisms by which soil carbon is lost to the atmosphere or transferred to stable carbon pools. USDA will also identify and quantify carbon sources, sinks and fluxes for all U.S. forest land, including marginal agricultural land and other potential conversion land-use types.

ARS will continue to focus on four broad research areas: 1) experimental determinations of the direct effects of rising atmospheric CO_2 levels, increasing temperatures, and their interaction with the physiology and performance of crop plants and with ecosystem processes that control productivity of grazing lands; 2) carbon and nitrogen cycling and fluxes between the terrestrial surface and the atmosphere, including sequestration of carbon in soils and vegetation; 3) changes in hydrological processes associated with climate change that may impact water quality, efficiency of use by crops, and availability for industry, urban use, and irrigated agriculture; and 4) the development of simulation models with required inputs for predicting responses of crops, watersheds, and managed ecosystems to global change.

CSREES will continue to support the USDA UV-B Monitoring Network. Information from this research network is combined with satellite-based measurements to provide an accurate climatological UV-B irradiance database. This database documents long-term trends and supports research and assessment of the potential for damage to ecosystems. Global Change research in CREES's National Research Initiative (NRI) Competitive Grants Program aims to increase understanding of the possible impacts of global environmental change on the sustainability of agriculture and forestry.

ERS will continues its analysis of the agricultural links to biodiversity, land-use change, and the ability to satisfy increased demands for agricultural goods and services while minimizing damage to the world's natural resources. Research will also assess the potential farm sector impacts of changing weather variability and farmer adaptation to changing environmental conditions.

FS global change research focuses on determining how atmospheric changes and potential climatic change may affect forest productivity, forest health, and species distributions. Ecosystem-scale experiments involving increased CO_2 and other environmental factors have begun at several sites representing major U.S. forest types. As the uncertainty in model predictions is reduced, analysts are describing likely socioeconomic effects of global change on forests in the various regions of the U.S. In FY 2001 the Forest Service will enhance its long term research on forest and grassland carbon cycles, with particular emphasis on the soil component. This enhanced carbon research effort will result in better information for use by forest resource managers to improve carbon cycle management on their lands.

NRCS will collect data necessary to build validated, verified baseline soil carbon inventories and assess policy driven impacts on soil carbon stocks at national, regional, and field-level scales. NRCS goals include establishing baseline soil carbon levels under various covers/management systems; developing a "use-dependent" soil carbon database integrated with national soils databases; collecting soil carbon data on a sample-based inventory frame for national and regional level inventory estimation; and testing the use of models and field collection of soil carbon data.

Related Research. In addition to focused USGCRP research, the USDA sponsors research contributing to the assessment of global change effects on the agricultural food and fiber production systems and the forest and forest ecosystems of the U.S. and worldwide. Programs include long-term studies addressing the structure, function, and management of forest and grassland ecosystems; research in applied sciences, including soils, climate, food and fiber crops, pest management, forest fish and wildlife, and social sciences; implementation of ecosystem management on the national forests and grasslands; and human interaction with natural resources.

Mapping of Budget Request to Appropriations Legislation. In the Agriculture, Rural Development, Food and Drug Administration, and Related Agencies Appropriations Bill, USGCRP activities are funded under Title I-Agricultural Programs, within the Agricultural Research Service (ARS), Cooperative State Research, Education, and Extension Service (CSREES) Research and Education Activities, and Economic Research Service (ERS) accounts; and under Title II-Conservation Programs, within the Natural Resources Conservation Service (NRCS) Conservation Operations account. In the Interior and Related Agencies Appropriations Bill, USDA USGCRP activities are funded in the USDA Forest Service (FS) section under Title II-Related Agencies, within the FS Forest Research account.

Department of Commerce / National Oceanic and Atmospheric Administration

DOC	Program Title	FY99	FY00	FY01 Request
NOAA	Aerosols Project	1.1	1.1	1.1
NOAA	Applications of Regional Forecasts	3.5	3.5	3.5
NOAA	Atmospheric Chemistry Project	6.6	6.6	6.6
NOAA	Carbon Cycle Science	3.9	4.9	4.9
NOAA	Climate Change Data and Detection	4.6	4.6	4.6
NOAA	Climate Dynamics and Experimental Prediction	16.5	18.2	18.5
NOAA	Climate Variability (CLIVAR)	17.0	18.0	18.0
NOAA	Human Dimensions of Global Change Research	1.7	1.7	1.7
NOAA	Global Energy and Water Cycle Experiment (GEWEX)	5.0	5.0	5.0
NOAA	Health of the Atmosphere*	0.8	0.8	0.8
NOAA	Paleoclimatology	2.3	2.3	2.3
NOAA	Climate Observations and Services**	0	0	28.0
	DOC Total	**63.0**	**66.7**	
	President's Request			**95.0**

* Not formally part of the NOAA Climate and Global Change Program; funding transferred to the NOAA Health of the Atmosphere Program.

** The NOAA USGCRP budget is now a combination of the Climate and Global Change Program and Climate Observations and Services; the requested FY 01 increase is directed almost entirely toward Climate Observations and Services.

Areas of Global Change Research. NOAA's global change efforts are designed to provide a predictive understanding of the climate system and its modes of variability, and to advance the application of this information in climate-sensitive sectors through a suite of process research, observations and modeling, and application and assessment activities. Specifically, NOAA's research program includes ongoing efforts in operational in situ and satellite observations with an emphasis on oceanic and atmospheric dynamics, circulation, and chemistry; understanding and predicting ocean-land-atmosphere interactions, the global hydrological cycle, and the role of global transfers of carbon dioxide among the atmosphere, ocean and terrestrial biosphere in climate change; improvements in climate modeling, prediction, and information management capabilities; the projection and assessment of variability across multiple timescales; the study of the relationship between the natural climate system and society and the development of methodologies for applying climate information to problems of social and economic consequences; and archiving, management, and dissemination of data and information useful for global change research.

FY 2001 Program Highlights. In FY 2001, NOAA will continue to advance understanding of: 1) whole-system dynamics and modes of climate variability, for example the El Niño-Southern Oscillation (ENSO) and the North Atlantic Oscillation (NAO); and 2) the application of information generated by this research to decision-making processes in climate-sensitive regions and sectors, such as agriculture, water management, hydropower, human health, and transportation infrastructure. In addition, NOAA will launch a new Climate Observations and Services program that will emphasize the transition of research data, observing systems, and understanding from experiments to applications, and from basic science to practical products. In FY 2001 NOAA will focus on climate monitoring capabilities and will: 1) begin deployment of a Climate Reference Network to monitor precisely fundamental variables such as temperature and precipitation across the U.S.; 2) implement critical upgrades and enhanced measurement programs at NOAA's atmospheric baseline observatories; 3) extend the global ocean observational networks to provide data critical for the modeling of seasonal climate variations such as El Niño and the modeling of long-term climate variations; and 4) improve climate data information access and products.

FY 2001 program highlights include the following:

- Improving regional-scale modeling and the prediction of seasonal to interannual variability over North America.
- Increasing understanding of the role of the Atlantic Ocean in climate changes, with an initial focus on the relationship between tropical Atlantic variability and the North Atlantic Oscillation.
- Continuing the advancement of a sustained Atlantic observing system to support CLIVAR research.
- Advancing the improvement of models and modeling systems for seasonal to interannual climate prediction and the ability to provide regional-scale forecasts and predicted probabilities of extreme events.
- Developing a comprehensive understanding of the effects of land-surface forcing on climate during the full annual cycle and the effects of orography on precipitation and water supply in the Missouri River Basin.
- Advancing detailed studies of past climate variability on seasonal to centennial timescales using century to millennia-long paleoenvironmental proxy records in order to improve the current understanding of seasonal to interannual variability.
- Developing and applying advanced statistical techniques to detect climate change signals and attribute these to specific causes.
- Increasing our understanding of the global transfers of CO_2 between the atmosphere, ocean, and terrestrial biosphere, thereby helping to constrain predictions of the uptake of anthropogenically-released CO_2 within these reservoirs, with an initial focus on the sinks of carbon in the North American continental region.
- Advancing efforts to reduce uncertainties in the understanding of direct radiative forcing by tropospheric aerosols through an integrated program focused on targeted, in-situ measurements of aerosols, integrated with model analyses.
- Characterizing the "ozone-friendliness" of substitutes for ozone-depleting gases, developing methods for the detection of the recovery of the ozone layer, and characterizing the regional variance of tropospheric ozone and its role in the heat budget.
- Advancing our understanding of societal vulnerability and current coping mechanisms related to climate variability on seasonal up to decadal timescales (including climate extremes and surprises), and the potential use of climate information for planning purposes.

- Advancing existing efforts to foster the application of forecast information in climate-sensitive regions and sectors such as agriculture, water management, energy, human health, and transportation infrastructure.

Related Research. In addition to focused USGCRP research, related activities include advance short-term weather forecasting and warning services; marine ecosystem research; prediction and observation systems in support of weather and seasonal to interannual climate forecasts; and facilitating the dissemination of global change information.

Mapping of Budget Request to Appropriations Legislation: In the Departments of Commerce, Justice, and State, the Judiciary, and Related Agencies Appropriations Bill, NOAA activities are funded under Title II–Department of Commerce and Related Agencies, within the NOAA Operations, Research, and Facilities account. In Appropriations Committee reports, funding for NOAA's USGCRP activities is the Climate and Global Change line-item within Oceanic and Atmospheric Research.

Department of Defense

Areas of Global Change Research. The Department of Defense does not support dedicated Global Change Research, but continues a history of participation in the USGCRP through sponsored research that concurrently satisfies national security requirements and stated goals of the USGCRP. A non-inclusive summary of unique Defense research and infrastructure associated with the USGCRP is described below. Because these efforts are not part of the formal USGCRP budget crosscut, a budget table is not included.

FY 2001 Highlights. This summary of Defense research and infrastructure describes DoD contributions to the USGCRP in the following three areas:

1) Observations. Ocean Acoustic Observatories, now in place and programmed, employ acoustic thermometry techniques to generate ocean temperature profiles along basin-scale source-receiver transects.

The North Pacific Acoustic Laboratory project capitalizes on networks previously installed by the Acoustic Thermometry of Ocean Climate (ATOC) project and extant Navy SOSUS arrays. Estimates of basin-scale sound speed (temperature) combined with altimetric and other data types clearly demonstrate the potential to improve understanding of gyre-scale ocean variability on seasonal and longer time scales.

The Arctic Climate Observations using Underwater Sound (ACOUS) project is a similar ocean thermometry project operating in the Arctic Ocean. This unique Navy program is included in the framework of the U.S.- Russia Joint Commission on Economic and Technical Cooperation (nee Gore-Primakov Commission).

Under the aegis of the multi-agency National Oceanographic Partnership Program (NOPP), a Pilot Project to assess the value, cost, and options to provide access to declassified Navy SOSUS data for civil research and education activities is planned for FY 2000-2002.

A unique program to use systems analyses techniques to develop optimal observing strategies is planned. This approach will use in-situ and remotely sensed data as well as synthetic data bases to construct optimal data collection schemes and data fields for operational and research models.

Navy continues to support the International Arctic Buoy Programme (IABP). The data and other products of the IABP are available on the World Wide Web at *http://iabp.apl.washington.edu*.

The Memorandum of Agreement between Navy and NSF was revised to continue Arctic research opportunities using available nuclear submarine platforms on a not-to-interfere basis.

A long term U.S. Research Vessel Fleet Plan is currently under development by the Office of Naval Research, NSF, and NOAA.

2) Data Assimilation. Several research efforts are underway to develop optimal, yet computationally efficient, methods to directly assimilate in-situ, remotely sensed, and synthetic observations/data into operational and research models. Initial efforts with satellite altimetry have been successful. Another effort is developing a relocatable, three-dimensional, multi-variate data assimilation scheme for ocean and atmosphere models. Extending this latter approach to fine-scale, multi-layer, ocean-atmosphere-coupled models is planned.

New computer code to process daily solar magnetic field data was transitioned to the Kitts Peak, Arizona, National Solar Observatory and installed as a regular operating procedure. During the current solar maximum, this data assimilation scheme will allow a superior description/prediction of space weather systems.

The Navy Fleet Numerical Meteorology and Oceanography Center is operating the GODAE World Wide Web server in return for near real time ocean drifter buoy data. This is a win-win event for the research and operational communities.

3) Modeling. The Navy's Polar Ice Prediction System (PIPS) continues to be improved. A fully operational 9-kilometer resolution coupled model employing multi-level thermodynamics over an expanded domain should be operational in 2001. This National model is expected to provide high-resolution operational forecasts of sea-ice extent and internal Arctic ice pack features, and generate realistic model initialization and boundary conditions for climate modeling. Results from the recently completed multi-agency Surface Heat Budget of the Arctic (SHEBA) project were incorporated in the PIPS model upgrade as well as other ocean and atmosphere modeling research.

The Defense Modeling and Simulation Office (DMSO) World Wide Web site *http://mel.dmso.mil* provides access to a variety of environmental and geospatial data and models. The multi-agency MEDEA group will continue to bridge the national security and civil community for access to classified environmental data.

All data and research results are routinely made available to the civil science community.

Related Research and Infrastructure. Other DoD-sponsored research and supporting infrastructure, not described above, also contribute to observing, understanding, and predicting environmental processes related to global change. Associated research programs include theoretical studies and observations of solar phenomena, monitoring and modeling of unique features in the middle and upper atmosphere, terrestrial and marine environmental quality research, and energy conservation measures.

DoD's continued investment in environmental infrastructure, such as the Oceanographic Research Vessel Fleet, the Cold Regions Research and Engineering Laboratory and the various services' operational oceanographic and meteorological computational centers, will continue to provide data and services useful to the USGCRP.

Mapping of Budget Request to Appropriations Legislation. In the Department of Defense Appropriations Bill, research associated with the USGCRP is funded under Title 1V – Research, Development, Test and Evaluation. In Appropriations Committee reports, nearly all funding is included within the budget for Defense Research Sciences.

Department of Energy

DOE	Program Title	FY 99	FY00	FY01 Request
BER	Climate & Hydrology	65.8	67.2	72.3
BER	Atmospheric Chemistry & Carbon Cycle	24.0	26.8	26.9
BER	Ecological Processes	13.7	11.9	12.0
BER	Human Interactions	8.1	8.2	8.2
BER	Small Business Innovative Research/ Technology Transfer (SBIR/STTR)	—	3.1	3.1
	DOE Total	**111.6***	**117.2**	
	President's Request			**122.5**

* Excludes $2.8M which was transferred to the SBIR program and $0.2M which was transferred to the STTR program in FY 1999.

BER Biological and Environmental Research Program

Areas of Global Change Research: Research supported by DOE's Office of Biological and Environmental Research (BER) addresses the effects of energy production and use on the global Earth system, primarily through studies of climate response. It includes research in climate modeling, atmospheric chemistry and transport, atmospheric properties and processes affecting the Earth's radiation balance, and sources and sinks of energy-related greenhouse gases (primarily CO_2). It also includes research on the consequences of atmospheric and climatic changes on ecological systems and resources, critical data needs for the detection and attribution of climate change, and tools and methods needed to conduct scientific assessments of climate change, and education and training of scientists and researchers in global change.

FY 2001 Program Highlights: The DOE Biological and Environmental Research program utilizes the unique, multidisciplinary facilities and capabilities of the DOE National Laboratories. BER supports global change research at these and other public and private research institutions, including universities. In FY 2001, DOE along with the other USGCRP agencies will continue to integrate the frontiers of climate and computational science to accelerate progress in climate simulation model development, testing, and application. In support of the USGCRP, highlights of the BER global change program includes activities in the following four key areas:

- **Climate and Hydrology:** DOE will continue observational and analytical research to acquire and interpret the data needed to describe the radiation balance from the surface of the Earth to the top of the atmosphere, to determine the atmospheric characteristics responsible for this balance, to improve the parameterization of the formation and evolution of clouds in climate models, and to see this understanding reflected in the improvement of climate models. This includes maintaining and operating Cloud and Radiation Testbed facilities in the central Great Plains, the north slope of Alaska, and in the Western Tropical Pacific. Additional new resources are requested by DOE in FY 2001 for climate modeling to substantially reduce the uncertainties in decade-to-century model-based projections of climate change and to increase the availability and utility of climate change simulations to the broader climate research and assessment communities.

- **Atmospheric Chemistry and Carbon Cycle:** DOE will continue field research and modeling activities to understand and document the net exchange of carbon dioxide between major terrestrial ecosystems and the atmosphere, to identify the biophysical processes controlling the net exchange, and to develop and test carbon cycle models for use in simulating the exchange of carbon between terrestrial ecosystems and the atmosphere. Support for an expanded AmeriFlux network of CO_2 flux measurement sites and for process studies at these sites will be a priority. In the Atmospheric Science area, research will focus on acquiring the data to understand the atmospheric processes that control the transport, transformation, and fate of energy-related chemicals and particulate matter. Field and laboratory studies will be supported in both atmospheric chemistry and environmental meteorology to acquire the data needed to develop and test models used to simulate and predict the transport, transformation, and fate of energy-related emissions in the atmosphere.

- **Ecological Processes:** In FY 2001, the DOE global change program will continue to support long-term experimental studies using, for example, the Free-Air CO_2 Enrichment (FACE) facilities to document the response of intact terrestrial ecosystems to alterations in climate and atmospheric composition, especially increasing concentrations of CO_2.

- **Human Dimensions:** The focus of research by DOE in this area in FY 2001 will be on developing and improving methods and tools for use in conducting integrated analyses of the climate change system from emissions of aerosols and greenhouse gases to the consequences of these emissions on climate and the resulting effects of human-induced climate change on economic and social systems. The research is intended to fill critical gaps in current integrated assessment modeling, including modeling of technology innovation and diffusion for the prediction of greenhouse emissions over long time scales. It also includes research to develop metrics and measures of economic costs of climate change for the purpose of comparing alternative policies or options for coping with or mitigating climate change. Finally, research will be supported on autonomous adaptation, i.e., that which may occur naturally in, for example, unmanaged ecosystems, or adaptation taken by individuals in response to actual or perceived climate change and to understand individual adaptation alternatives.

Related Research: DOE plays a major role in the President's Climate Change Action Plan to reduce greenhouse gas emissions through changes in energy supply and improvements in energy efficiency and conservation. Although it builds on but is not part of the USGCRP, research to understand how to enhance carbon sequestration in terrestrial and marine ecosystems and the potential environmental consequences of enhancing the sequestration in these systems are part of the Climate Change Technology Initiative.

Mapping of Budget Request to Appropriations Legislation: In the Energy and Water Development Appropriations Bill, DOE USGCRP activities are funded under Title III, Department of Energy, within the Energy Supply, Research, and Development Activities account. In Appropriations Committee reports, funding for DOE's USGCRP programs is included within the Biological and Environmental Research account.

Department of Health and Human Services/ National Institutes of Health

HHS/NIH	Program Title	FY99	FY00	FY01 Request
NIEHS	Human Health Effects of Exposure to UV Radiation and CFC Replacement Chemicals	5.0	5.4	5.6
NEI	Health Effects of UV Radiation	10.7	12.2	12.8
NCI	Health Effects of UV Radiation	24.3	27.8	29.5
NIAMS	Health Effects of UV Radiation	0.0	0.3	0.3
	HHS/NIH Total	**39.9**	**45.7**	
	President's Request			**48.1**

NIEHS	National Institute of Environmental Health Sciences
NCI	National Cancer Institute
NEI	National Eye Institute
NIAMS	National Institute of Arthritis and Musculoskeletal and Skin Diseases

Areas of Global Change Research. Four NIH institutes support research on the health effects of UV and near-UV radiation. Their principal objectives include an increased understanding of the effects of UV and near-UV radiation exposure on target organs (e.g., eyes, skin, immune system) and of the molecular changes that lead to these effects, and the development of strategies to prevent the initiation or promotion of disease before it is clinically defined. In addition, NIEHS supports research on the health effects of CFC replacement chemicals, including studies on the metabolism and toxicity of HCFCs and halogenated hydrocarbons.

FY 2001 Program Highlights. The NIEHS program supports grants and intramural projects that investigate the effects of UV exposure on the immune system, aging process, sensitive tissues such as the retina and skin, and methods to reduce these harmful effects. Other projects involve the comparison of mutagenic potential in bacteria of UV and near-UV radiation at levels found in natural sunlight and at levels anticipated with a 15 percent depletion of stratospheric ozone. Several projects supported by NIEHS are investigating molecular changes in DNA that lead to aberrations and mutations in human tissue, rodents, fruit flies, and bacteria, and the variety of ways these organisms repair damage to DNA resulting from UV exposure.

A major NEI initiative is underway to determine how and why eye cataract develops and to search for ways to prevent or slow the progression of cataract, an age-related eye disease that affects 17-20 million people globally. This project is investigating the role of UVB radiation, which has been implicated as a specific risk factor in cataract development. Another important area of NEI research is the understanding of certain detoxification systems in the eye and how they combat damage from UVB radiation. The goal of this effort is to identify drugs that might have therapeutic or preventative applications.

The NCI is supporting a wide range of studies to characterize the etiology, biology, immunology, and pathology of a variety of changes in the skin, including photoaging, non-melanoma skin cancers, and melanoma caused by exposure to UV radiation. Other research is exploring UV-induced immunosuppression, which is critical to the development of UV-induced skin tumors, and the cellular and molecular basis for the genetic predisposition to UVB-induced skin cancer in people with Basal Cell Nevus Syndrome.

NIAMS supports basic and clinical research on the effect of UVA and UVB radiation on skin.

Related Research. In addition to research areas that are designated as part of the USGCRP, NIEHS conducts research related to other impacts of global change on human health, including the effects of environmental and occupational exposures to air pollution, agricultural chemicals, and materials used in technologies to mitigate or adapt to climate change. Exposures of special concern for FY 2001 include those that contribute to the greatly increased incidence of childhood asthma and that disrupt the normal functioning of the endocrine system. Renewed concern about emerging and reemerging infectious diseases has prompted increased attention to a variety of diseases whose incidence would be affected by environmental change. Other HHS agencies provide significant resources for research on the prevention of and treatment for water-, food- and vector-borne diseases, such as cholera, salmonella, encephalitis, malaria, dengue, and Lyme disease.

Mapping of Budget Request to Appropriations Legislation. In the Departments of Labor, Health and Human Services, and Education and Related Agencies Appropriations Bill, USGCRP activities are funded under the NIH section of Title II–Department of Health and Human Services.

Department of the Interior

DOI	Program Title	FY99	FY00	FY01 Request
USGS	Global Change Research	26.7	27.0	25.4
	DOI TOTAL	**26.7**	**27.0**	
	President's Request			**25.4**

USGS U.S. Geological Survey

Areas of Global Change Research. Research at the Department of the Interior's U.S. Geological Survey (USGS) contributes directly to the USGCRP's intellectual framework of a whole-system understanding of global change (i.e., the interrelationships among climate, ecological systems, and human behavior). The USGS examines terrestrial and marine processes and the natural history of global change, including the interactions between climate and the hydrologic system. Studies seek to understand the character of past and present environments and the geological, biological, hydrological, and geochemical processes involved in environmental change.

FY 2001 Program Highlights. In FY 2001, the USGS will support ongoing efforts across a broad area of global change research, with a focus on understanding the sensitivity of natural systems and impacts of climate change and variability, surficial processes, and other global change phenomena on the Nation's lands and environments at the regional scale. Specific goals of the program are: to improve the utility of global change research results to land management agencies; to emphasize monitoring the landscape and developing technical approaches to identifying and analyzing changes that will take advantage of a burgeoning archive of remotely sensed and in situ data; and to emphasize the response of biogeographic regions and features, particularly montane, coastal, and inland wetland ecosystems. In direct support of USGCRP, the USGS includes activities in the following areas:

Biogeochemical Cycling - Research is developing an understanding of the exchanges of water, energy, and nutrients between the atmosphere and land surface. The processes that control the cycling and fate of carbon and other nutrients in soils, rivers, lakes, reservoirs, and estuarine systems are critical to understanding issues related to erosion, sediment transport, biogeochemical budgets, snowpack chemistry, surface hydrology, and climate response.

Carbon Cycle - Studies are developing a quantitative understanding of the role of land-use change and associated erosion and sedimentation processes on carbon storage and nutrient cycles within the Mississippi Basin. Rates of organic carbon accumulation, erosion, and burial are used to develop whole-basin models of these dynamic relationships.

Climate History - Climate history research focuses on understanding the rates and magnitudes of decadal to millennial scale natural changes in climate and determining how those changes have affected the environment.

Hydroclimatology - This research consists of monitoring trends in the accumulation and dissipation of snow and ice stored in selected U.S. benchmark glaciers.

Impacts on Terrestrial and Coastal Ecosystems, Coastal Wetlands, and Fish and Wildlife – This research determines the sensitivity and response of natural systems and ecological processes to multiple environmental factors, including existing climate and natural and anthropogenic impacts, at the local, landscape, regional, and continental level. It provides the scientific knowledge and technologies for conservation, rehabilitation, and management of sustainable ecosystems needed by land management agencies of the Federal and state governments.

Land Surface Characterization – This area includes research and development of techniques to monitor, analyze, describe, apply, and predict land use, land cover, and other surface characteristics data. Data sets are used to characterize and map the Earth's surface, model land surface processes, detect changes over time, and forecast the response of the land surface to changes in climate, environment, land use, and land cover.

Satellite Data Management and Dissemination - The USGS continues to operate and enhance the capabilities of the EROS Data Center to serve as the National Satellite Land Remote Sensing Data Archive, by maintaining existing datasets, adding new data sets, and converting older data sets from deteriorating media to modern, stable media.

Terrestrial Earth Surface Processes - Research examines the impact of climatic variability and change on earth surface processes, including vegetation change, soil and sediment dynamics, and carbon sequestration. A detailed history of vegetation change in the western U.S. and southern Alaska is being constructed.

Related Research. DOI also sponsors contributing research programs addressing the collection, maintenance, analysis, and interpretation of short- and long-term land, water, biological, and other geological and biological processes and resources through dispersed observing networks; research in land use and land cover, including creation of maps and digital data products; and inventorying and monitoring of biological habitats, resources, and diversity.

Mapping of Budget Request to Appropriations Legislation. In the Interior and Related Agencies Appropriations Bill, DOI USGCRP activities are funded under Title I-Department of the Interior. Funding for U.S. Geological Survey USGCRP programs is included within the USGS Survey, Investigations, and Research account.

Environmental Protection Agency

EPA	Program Title	FY99	FY00 Enacted	FY01 Request
ORD	Assessment of Consequences of Climate Variability and Change	16.0	17.6	19.7
ORD	Biology and Biogeochemistry of Ecosystems	0.0	3.0	3.0
	EPA Total	**16.0**	**20.6**	
	President's Request			**22.7**

ORD Office of Research and Development

Areas of Global Change Research. EPA's Global Change Research Program is an assessment-oriented program with primary emphasis on understanding the potential consequences of climate variability and change on human health, ecosystems, and socioeconomic systems in the United States. This entails: (1) improving the scientific basis for evaluating effects of global change in the context of other stressors and human dimensions (as humans are catalysts of and respond to global change); (2) conducting assessments of the risks and opportunities presented by global change; and (3) assessing adaptation options to improve society's ability to effectively respond to the risks and opportunities presented by global change as they emerge.

FY 2001 Program Highlights. The program has made a major commitment to the National Assessment activities organized through the USGCRP. The Global Change Research Act of 1990 mandates that the USGCRP conduct periodic assessments of the potential consequences of global change for the United States. (These periodic assessments are to be conducted not less than every four years.) As a member of the USGCRP, EPA's Global Program will continue to make significant contributions to the ongoing U.S. National Assessment Process. The EPA-sponsored assessments will continue to be conducted through public-private partnerships that actively engage researchers from the academic community, decision makers, resource managers, and other affected stakeholders in the assessment process.

EPA's intramural assessment program has four areas of emphasis: (1) human health; (2) air quality; (3) water quality; and (4) ecosystem health. These four focus areas are consistent with EPA's mission and the strengths of EPA's research program.

The first of four focus areas is Human Health. Because health is affected by a variety of social, economic, political, environmental, and technological factors, assessing the health impacts of global change is a complex challenge. As a result, health assessments in EPA's Global Program go beyond basic epidemiological research to develop integrated health assessment frameworks that consider the effects of multiple stresses, their interactions, and human adaptive responses. Along with health sector assessments conducted in conjunction with the USGCRP National Assessment process, there are research and assessment activities focused on the consequences of global change on weather-related morbidity and vector- and water-borne diseases. In addition, the results from the Global Program's air quality assessments will be used to evaluate health consequences. In FY 2001, the program's focus will be on the assessment of the effects of climate change on weather-related morbidity.

The second focus area is Ecosystems. The EPA's mission is not only to protect human health but also to safeguard the natural environment. EPA has pledged to provide environmental protection that "contributes to making communities and ecosystems diverse, sustainable, and economically productive." Consistent with this goal, EPA's Global Program is considering comprehensive ecosystem issues related to global change. Three research and assessment activities are planned that evaluate the effects of global change on 1) aquatic ecosystems (which may include lakes, rivers, and streams; wetlands; and estuaries and coastal ecosystems); 2) invasive non-indigenous species; and 3) ecosystem services. The assessment of aquatic ecosystems will contribute to water quality assessments of pollutants and pathogens and of biocriteria. The ecosystem services assessment will draw on work from the other ecosystem assessments. In FY 2001, the program's focus will be on the assessment of the effects of global change on aquatic ecosystems.

The third focus area is Air Quality. Few studies have investigated the effect of global change on air quality. Given EPA's legal mandates with respect to air pollution and substantial capability and expertise in modeling air quality and evaluating integrated response actions, examining the effects of global change on air quality is a logical focus of the Global Program. Assessments are planned that will examine the potential consequences of global change on tropospheric ozone and particulate matter. Each of these assessments is paired with a related human health assessment. In FY 2001, the program's focus will be on the assessment of the effects of global change on tropospheric ozone.

The fourth focus area is Water Quality. Water quality is affected by changes in runoff following changes in precipitation and evapotranspiration and/or changes in land use. The program plans two assessments of the possible impacts of global change (climate and land use change) on water quality. Both water quality assessments will either contribute to or benefit from human health and ecosystems assessments. In addition, results from the assessment of pollutants and microbial pathogens will be used in the assessment of biocriteria.

Intramural and extramural research contribute to all of EPA's assessments. In an attempt to capitalize on expertise in the academic community, a significant portion of the program's resources are dedicated to extramural research grants administered through the STAR (Science to Achieve Results) grants program. The STAR program focuses on two principal areas related to global change research — science to support assessments of consequences of global change and human dimensions research.

Related Research. In addition to the focused USGCRP activities, EPA conducts research that contributes to the characterization and understanding of risks to ecosystems and to human health. The ecosystem-based research is designed to understand and predict ecosystem exposure, responses, and vulnerabilities to high-risk chemicals and non-chemical (e.g., invasive species, genetically altered organisms) stressors at multiple scales of biological organization and geographic scales. The research in human health is oriented towards assessing the cumulative health risks (e.g., cancer, reproductive, cardiovascular) to humans, including high-risk subpopulations (e.g., children), from chemical stressors emanating from multiple sources. Both of these major research areas will be impacted by and are inextricably interrelated with climate change.

Mapping of Budget Request to Appropriations Legislation. In the Departments of Veterans Affairs and Housing and Urban Development, and Independent Agencies Appropriations Bill, EPA USGCRP activities are funded under the EPA section of Title III – Independent Agencies, within the Science and Technology account. Appropriations Committee report language may specify more directly the funding for global change research.

National Aeronautics and Space Administration

NASA Program Title	FY 1999 Enacted	FY 2000 Planned	FY 2001 Request
Atmospheric Chemical Modeling	7.2	7.0	7.0
Atmospheric Dynamics & Remote Sensing	5.5	5.1	5.1
Biological Oceanography	4.4	4.7	4.7
Ecological Processes	16.9	16.9	16.9
EOS Science	46.4	55.0	52.2
Global Data Integration & Validation	3.4	3.7	3.7
Global Modeling and Analysis Program	7.7	7.3	7.3
GLOBE	5.0	5.0	5.0
Interdisciplinary Research and Analysis	15.7	16.4	20.7
Global Water Cycle	0.0	0.0	5.0
Global Carbon Cycle	0.0	0.0	5.0
Global Ocean Research (incl. NOPP)	0.0	1.0	3.0
Land Cover and Use Change	5.0	6.3	6.3
Land Surface Hydrology/water cycle	5.8	6.3	6.3
Mission Analysis Program	40.6	41.9	50.0
Ocean Color Data Purchase/Sea WIFS	2.5	2.6	2.7
Pathfinder Science Studies	3.5	3.5	0.0
Physical Oceanography & Ocean Modeling	8.5	8.5	8.5
Polar Programs	5.4	6.5	6.5
Radiation Science Program	7.7	9.9	10.9
Stratospheric Chemistry	17.4	19.2	16.7
Troposheric Chemistry	9.0	9.0	9.0
NASA Global Change Science Program	**217.5**	**235.8**	**252.4**
Earth System Science Pathfinder	62.3	94.5	105.9
EOS Data and Information Systems (EOS DIS)	261.7	261.9	252.0
EOS Flight Development (less Tech Inf.)	410.5	370.8	397.9
EOS Special Spacecraft	116.0	120.4	86.7
Information Systems	6.1	7.6	9.7
LANDSAT	17.0	9.6	1.4
Launch Services	4.2	0.0	0.0
Mission Operations	49.6	47.6	42.7
Total Ozone Mapping Spectrometer (TOMS)	9.9	24.8	0.5
NASA Global Change Hardware Development	**937.3**	**937.2**	**896.8**
NASA USGCRP Budget	**1,154.8**	**1,173.0**	**1,149.2**

Areas of Global Change Research. NASA research efforts in global change involve space-based studies of the Earth as an integrated system, including research and satellite programs studying atmospheric chemistry and ozone; ocean surface winds, tropical precipitation and the global hydrological cycle and climate variability cycle; and the global carbon cycle, ocean biological productivity and land surface vegetation and ecosystems.

The space-based activity complements ongoing ground-based research programs in the observation, understanding, and modeling of radiation, climate dynamics, and hydrology and water resources; ecosystem dynamics and biogeochemical cycles; atmospheric chemistry; and the processing, archiving, retrieval, dissemination, and use of global change data. The focus is Earth system science, which involves interdisciplinary research and coupled modeling. Development of algorithms for retrieval of the information content of space-based, remotely-sensed observations is carried out as part of the flight mission.

FY 2001 Program Highlights. The overall goal of NASA's Earth Science Enterprise (ESE) is to understand the Earth system and the effects of natural and human-induced changes on the global environment. To preserve and improve the Earth's environment for future generations, policies and decisions worldwide should have the strongest possible scientific basis. The vantage point of space provides information that is obtainable in no other way about the Earth's land, atmosphere, ice, oceans, and biota, as well as the impact of humans on the Earth system.

The science and observations of ESE are becoming increasingly important as nations work to meet the demand for economic progress by a growing global population. In addition, remote sensing has the potential to improve dramatically crop and forest yield predictions, seasonal and interannual climate forecasts, urban planning, mineral exploration, fisheries management, and many other activities of socioeconomic importance. In concert with the global change research community, the ESE is utilizing the vantage point of space to lead the development of knowledge required to support the complex national and international policy decisions that lie ahead.

This edition of *Our Changing Planet* continues to divide the ESE budget into two main components: (1) scientific research; and (2) the budget associated with satellite, aircraft, and balloon measurements, operations, and data processing and distribution (including mission costs such as launch, flight, instrument and technology development, fabrication assembly, integration, and testing, as well as mission operation support).

Scientific Research. The scientific research component of the ESE budget is supported by an integrated science plan that relates research plans to space observations, and fully integrates the Earth Observing System (EOS) and non-EOS science. EOS is a program of multiple spacecraft and interdisciplinary science investigations, designed to provide a 15-year data set of key parameters needed in order to understand global climate change. The major themes of NASA's ESE Science Research Plan are consistent with the newly refined USGCRP Program Elements for FY 2000-2001.

Against the backdrop of the overall ESE effort to better understand the state and health of the Earth's life-support systems, NASA's FY 2001 research will target specific research issues important to national and international environmental and economic security. Through increases in Interdisciplinary Research and Analysis funding and targeted augmentations in the Radiation Science component of the Research & Analysis Program, NASA will participate in the new interagency Global Water and Energy Cycle

Science Program. New research will focus on exploiting data from new satellites (i.e., EOS Terra, Landsat 7, VCL, EO-1) to document the role of land-cover change, ecosystem disturbances, and interannual variability in terrestrial and marine ecosystem productivity in regional and global carbon dynamics. Another important priority is to provide an accurate assessment of the extent and health of the world's forests, grasslands, and agricultural resources.

In a time of rapid, and often unrecorded, land-use change, observations from space are the only global source of objective information on the human use of land. A related priority is to improve understanding and prediction of seasonal to interannual climate variation. Reducing uncertainties in climate predictions out to a season or a year in advance can help dramatically improve the efficiency of water use for agriculture and hydropower, as well as improve contingency planning for energy demand and in other economic sectors.

There is increasing evidence that predictions of extreme weather events can be improved by understanding their links to interannual climate phenomena, such as the El Niño events. The ESE Science Plan also calls for special attention to measuring and modeling the relative influence of forcing factors in long-term climate change, including clouds, aerosols, and greenhouse gases, in order to improve the understanding and prediction of climate on time scales of decades to centuries. A continuing priority area for ESE is to understand the causes and consequences of changes in atmospheric ozone and the nature and timing of the expected recovery of stratospheric ozone in an atmosphere with increased abundances of greenhouse gases. Research to resolve questions related to stratospheric ozone depletion continues to make great progress, and increased emphasis is now being focused on the changing composition of the lower atmosphere, which is especially sensitive to the unprecedented growth of pollutant emissions in East Asia and other rapidly developing regions.

Satellite, Aircraft, and Balloon Measurements, Operations, and Data Processing and Distribution. The Earth Observing System is a program of multiple spacecraft (the Terra, Aqua, and Aura series, Landsat-7, and others) and interdisciplinary science investigations to provide a 15-year data set of key parameters needed to gain a fuller understanding of global climate change.

Preceding EOS are a number of individual satellite and Shuttle-based missions which are helping to reveal the basic processes of: atmospheric chemistry (Upper Atmosphere Research Satellite-UARS/1991), ozone distribution and depletion (Total Ozone Mapping Spectrometer-TOMS/1978, 1991, 1996, and 2001), ocean topography and circulation (TOPEX/Poseidon/1992), ocean winds (NASA Scatterometer-NSCAT/1996; QuikSCAT/1999), ocean color (Sea Wide-Field-of-View Sensor-SeaWiFS/1997), and global tropical precipitation (Tropical Rainfall Measuring Mission-TRMM/1997), among others. These provide the scientific and technological foundation on which EOS builds. TRMM was launched in November 1997 and is still operating. It will provide important data on precipitation in the tropics that will help better understand the global hydrological cycle.

The first EOS satellite launches began in 1999, with Landsat-7, QuikSCAT, ACRIMsat and Terra. In February 2000, the Shuttle Radar Topography Mission successfully collected 30-m and 90-m horizontal resolution topographic data on the entire land surface of the Earth between 60°N and 56°S. In April 2000, science data from Terra began flowing to an enthusiastic science community. The balance of the calendar year 2000 should see the launches of QuikTOMS, SAGE III for atmospheric research, and Jason-1 for ocean altimetry. As ESE moves into 2001, data from Aqua will provided vastly improved measurements of atmospheric temperature and humidity, as well as complementing Terra's global biosphere and clouds and radiation measurements. 2001 will also see the launch of ICEsat and SORCE.

In addition, NASA will launch the New Millennium Program Earth Observer-1 technology demonstration mission in 2000, designed to make future Landsat-type missions possible at vastly reduced size and cost. The New Millennium Program (NMP) provides for the infusion of innovative new technologies into ESE, with an initial focus on the EOS follow-on missions, and will emphasize fast-track development and low-cost demonstration missions. These technologies, which will lead to the development of smaller and lighter-weight instruments, will reduce annual program expenditures in the post-2002 time frame.

Complementing EOS will be a series of small, rapid-development Earth System Science Pathfinder (ESSP) missions to study emerging science questions and make innovative measurements in parallel with the systematic, long-term measurements begun with EOS. ESSP will feature low life-cycle costs, peer-reviewed science, and missions based on best science value. The first two ESSP missions — Vegetation Canopy Lidar (VCL) and Gravity Recovery and Climate Experiment (GRACE) — were selected and are scheduled for launch in 2000 and 2001, respectively.

ESE has adopted an evolutionary approach to fulfilling its mission and goals. Future missions needed to achieve continuity for systematic measurements, together with those in the exploratory mode of ESSP, will be implemented according to the "better/faster/cheaper" paradigm. ESE will use commercially available spacecraft in a "catalog" procurement mode to reduce the cost and development time required to prepare a mission for launch. Meanwhile, ESE will invest upfront in instrument technology development, and base its mission selection on both scientific need and technology readiness. ESE has developed a Research Strategy for the new decade. It will serve as the basis for prioritizing and selecting satellite missions for the 2003-2010 time frame. This document is currently being reviewed by the National Academy of Sciences. Science needs for the new decade are already sufficiently mature to have initiated formulation of two key missions in this timeframe. The NPOESS Preparatory Project will serve to extend essential measurements from Terra and Aqua as well as demonstrate new instruments for the converged weather satellite program. A Landsat continuity mission, which may take the form of a commercial and/or international collaboration, will extend the valuable 30m global land cover data set beyond the lifetime of Landsat 7. Both missions are envisioned to launch in 2005.

Related Research. All NASA global change research is included in the USGCRP program.

Mapping of Budget Request to Appropriations Legislation. In the Departments of Veterans Affairs and Housing and Urban Development, and Independent Agencies Appropriations Bill, National Aeronautics and Space Administration USGCRP activities are funded under the NASA section of Title III-Independent Agencies, as part of the Science, Aeronautics, and Technology account. Within this account, Appropriations Committee reports specify funding for the Earth Science program.

National Science Foundation

NSF Program Title	FY99	FY00	FY01 Request
Antarctic Ecosystems	1.0	1.0	1.0
Arctic System Science (ARCSS)	13.8	14.3	14.3
Carbon Cycling	2.8	5.2	8.1
Climate Modeling, Analysis & Prediction (CMAP)	13.0	13.4	13.4
Climate Variability and Predictability (CLIVAR)	13.6	15.9	18.8
Coastal Long-Term Ecological Research (cLTER/LMER)	3.2	3.2	3.2
Earth System History	18.1	19.3	19.5
Ecological Diversity	6.2	6.7	6.7
Ecological Rates of Change (EROC)	3.2	3.2	3.2
Geodata	2.7	2.8	2.8
Global Ocean Ecosystems Dynamics (GLOBEC)	13.3	14.9	14.9
Global Tropospheric Chemistry Program (GTCP)	13.9	14.3	14.3
Greenhouse Gas Dynamics (GGD)	0.2	0.2	0.2
Human Dimensions of Global Change	13.6	14.0	14.0
Joint Global Ocean Flux Study	10.2	7.9	5.0
Methods and Models for Integrated Assessment	3.4	3.4	3.4
Ocean Observation, Data Assimilation, and Modeling (OODAM)	4.1	4.6	4.6
Polar Ozone Depletion/UV Radiation Effects	4.2	4.2	4.2
Regional Research Institutes	3.2	3.2	3.2
Ridge Interdisciplinary Global Experiments (RIDGE)	3.3	3.3	3.3
Sea Level Changes	6.2	6.4	6.2
Solar Influences	7.2	7.4	7.4
Water & Energy: Atmospheric, Vegetative & Earth Interactions	9.6	9.7	9.7
World Ocean Circulation Experiment (WOCE)	11.8	9.1	6.2
NSF Total	**181.7**	**187.5**	
President's Request			**187.5**

Areas of Global Change Research. NSF programs address global change issues through investments in challenging ideas, creative people, and effective tools. In particular, NSF global change research programs support research and related activities to advance the fundamental understanding of dynamic physical, biological, and socioeconomic systems and the interactions among them. The programs encourage interdisciplinary activities with particular focus on Earth system processes and the consequences of change. NSF programs facilitate data acquisition and information management activities necessary for fundamental research on global change, promote the enhancement of models designed to improve our understanding of Earth system processes and interactions, and to develop advanced analytic methods to facilitate basic research. NSF also supports fundamental research on the general processes used by organizations to identify and evaluate policies for mitigation, adaptation, and other responses to the challenge of varying environmental conditions.

FY 2001 Program Highlights. During FY 2001, NSF will continue to support research and related activities across its broad range of environmental programs. New NSF-wide initiatives will emphasize environmental processes, information technology, and human resource issues. As in preceding years, NSF will continue to invest in major international collaborative field, modeling, and analysis programs. Major emphasis will focus on: 1) Arctic systems; 2) climate modeling and predictability; 3) global carbon and global water cycles; 4) atmospheric chemistry analysis; 5) global ocean ecosystems; 5) Earth system history; and 6) human dimensions of climate change. In addition, the Ocean Observations, Data Assimilation, and Modeling (OODAM) and Climate Variability and Predictability (CLIVAR) programs will be expanded in FY 2001.

The Arctic systems program will support the Surface Heat Budget of the Arctic (SHEBA) Ocean project to incorporate sea-ice albedo and cloud-radiation feedback data in models that predict the impact of global warming on the fate of sea ice, studies of the impact of global change on terrestrial ecosystems, incorporation of atmospheric chemistry into the snow and ice record, and the processes influencing the marine biogeochemistry at the shelf-basin interface. A new initiative will seek to establish a center for the study of the Human Dimensions of the Arctic System.

Emphasis will continue to be placed on climate modeling activity at various academic and research centers. The centerpiece will be the Community Climate System Model (CCSM), the only climate model available for community use. CCSM is being developed, tested, and applied cooperatively by scientists from the National Atmospheric Research Center, universities, and private and government laboratories. Notably, it is the first model to simulate several centuries of Earth's climate without the need for artificial "flux corrections." In FY 2001, a second generation CCSM will be available and will include improved atmospheric and ocean physics, land surface characterizations, sea-ice processes, and biogeochemistry.

While the World Ocean Circulation Experiment and the Joint Global Ocean Flux Study are entering their synthesis phase, other NSF programs will take on increased emphasis. Two particular examples are the global carbon cycle and the global water cycle. A plan for an integrated carbon cycle science will address terrestrial and oceanic reservoirs and consider biophysical and ecological feedback mechanisms. A global water cycle science plan is nearing completion. It will focus on understanding the causes of water cycle variability, both natural and human-induced; improving predictions and understanding the impacts and relationships to biogeochemical cycles.

The field portion of the Indian Ocean Experiment (INDOEX) has now concluded. It investigated the transport of pollutants and how they affect atmospheric composition and solar radiation processes. In the future, analysis and modeling work will continue to focus on particles and their influence on surface temperatures which have important consequences for the region's climate.

Related Research. In addition to the research focused on global change, NSF will continue to conduct research on topics which are closely related to global change including laboratory and field studies of the atmosphere and ocean. These physical, chemical, geophysical, and biological investigations will provide additional understanding of Earth processes in support of the objectives of the USGCRP. In addition, many NSF-sponsored research projects consider interactions linking ecosystems and human activities with other factors including climate variability and change. Thus, much NSF research support may be considered "contributing research." Examples include the Long-Term Ecological Research sites, which provide insights into ecosystem responses and simultaneously perspectives on ecological responses to other stresses beside climate change. NSF supports research projects examining economic, cultural, and behavioral responses to global environmental change. Finally, the need to understand the responses of society to risks has fostered research in risk assessment and management.

Mapping of Budget Request to Appropriations Legislation. In the Departments of Veterans Affairs and Housing and Urban Development, and Independent Agencies Appropriations Bill, NSF USGCRP activities are funded under the NSF section of Title III – Independent Agencies within the NSF Research and Related Expenses account.

Smithsonian Institution

SI	Program Title	FY99	FY00	FY01 Request
NMNH/STRI	Long-Term Environmental Change	1.6	1.6	1.6
SAO/NASM/SERC	Monitoring Natural Environmental Change	1.2	1.2	1.2
NZP/NMNH/SERC	Biological Responses	4.2	4.2	4.2
	Smithsonian Total	**7.0**	**7.0**	
	President's Request			**7.0**

NASM	National Air and Space Museum
NMNH	National Museum of Natural History
NZP	National Zoological Park
SAO	Smithsonian Astrophysical Observatory
SERC	Smithsonian Environmental Research Center
STRI	Smithsonian Tropical Research Institute

Areas of Global Change Research. Within the Smithsonian Institution, global change research is conducted at the Smithsonian Astrophysical Observatory, the National Air and Space Museum, the Smithsonian Environmental Research Center, National Museum of Natural History, Smithsonian Tropical Research Institute and National Zoological Park. Research is organized around themes of atmospheric processes, ecosystem dynamics, observing natural and anthropogenic environmental change on daily to decadal time scales, and defining longer-term climate proxies present in the historical artifacts and records of the museums as well as in the geologic record at field sites. The Smithsonian Institution program strives to improve knowledge of the natural processes involved in global climate change, provide a long-term repository of climate-relevant research materials for present and future studies, and to bring this knowledge to various audiences, ranging from scholarly to lay public. The unique contribution of the Smithsonian Institution is a long-term perspective, e.g. undertaking investigations that may require extended study before producing useful results and conducting observations on sufficiently long (e.g. decadal) time-scales to resolve human-caused modification of natural variability.

FY 2001 Program Highlights. Researchers at SAO will study stratospheric trace species that play an important role in ozone photochemical cycles using balloons, airplanes, and satellites. Solar activity and irradiance are being studied to better understand the climatic effects of solar variability. Ongoing global sea-level change is being estimated using space geodetic measurements. Research at NASM emphasizes the use of remote-sensing data to improve theories of drought, sand mobility, soil stability, and climate change in the eastern Sahara.

Studies at NMNH and STRI focus on the paleoecology of climate change. At SERC, measurements will be made of spectral UV-B in Maryland (>25 y record), Florida, Arizona and other sites in the U.S. This data will be electronically disseminated to meet the needs for assessing the biological and chemical impact of varying UV exposure.

Several SI programs will examine biological responses to global change and increase public understanding of global change issues. At SERC, research will be conducted on the responses of global ecosystems to increasing CO_2, exotic species introductions, and solar UV. At STRI, research will be conducted on the effects of climate change (including CO_2 increase) on tropical ecosystems. Biodiversity education and research will be performed at STRI, NMNH and NZP. Tropical biodiversity research programs monitor global change effects through repeated sampling of flora and fauna in tropical forests, and identifying the physical and biological processes of growth and decline of species. Other studies on ecosystem response to increasing habitat fragmentation will be conducted at NZP.

The general public and research community will be informed of global change research conducted by Smithsonian and other USGCRP agencies via exhibits, such as the *"Forces of Change"* exhibit at NMNH, educational programs, and a global change information web page.

Related Research. Contributing activities include research conducted by several units within the Smithsonian in a variety of habitats concerning natural and human-induced variations in species, populations-communities and ecosystems. These studies help clarify the relative importance of global change effects as one of several agents of ecological change. Studies of environmental change over long time periods are aided by the Institution's collections. Utilized by staff and researchers from other institutions, these materials provide raw data for evaluating changes in the physical and biological environment that occurred before human influences.

Mapping of Budget Request to Appropriations Legislation. In the Interior and Related Agencies Appropriations Bill, Smithsonian Institution USGCRP activities are funded in the SI section of Title II—Related Agencies, within the Salaries and Expenses account. Appropriations Committee reports specify funding for a Sciences line item component of this account, which includes USGCRP programs.

CONTACT INFORMATION

U.S. Global Change Research Program Coordination Office
Richard Moss, Executive Director
400 Virginia Avenue, SW
Suite 750
Washington, DC 20024
202-488-8630 (voice)
202-488-8681 (fax)
rmoss@usgcrp.gov (e-mail)
http://www.usgcrp.gov/
http://www.gcdis.usgcrp.gov/

National Assessment Coordination Office
Michael MacCracken, Executive Director
400 Virginia Avenue, SW
Suite 750
Washington, DC 20024
202-488-8630 (voice)
202-488-8681 (fax)
mmaccrac@usgcrp.gov (e-mail)
http://www.usgcrp.gov/

Intergovernmental Panel on Climate Change U.S. Coordination Office
Neil Leary, Head
400 Virginia Avenue, SW
Suite 750
Washington, DC 20024
202-314-2225 (voice)
202-488-8678 (fax)
ipcc@usgcrp.gov (e-mail)
http://www.usgcrp.gov/ipcc

For additional information on USGCRP activities, or to obtain a copy of this document, contact the Global Change Research Information Office (GCRIO) at the following address:

P.O. Box 1000
61 Route 9W
Palisades, New York 10964 USA
+(1)(914) 365-8930 (voice)
+(1)(914) 365-8922 (fax)
help@gcrio.org (e-mail)
http://www.gcrio.org/

Abstract

Our Changing Planet: The FY 2001 Global Change Research Program is a report to Congress supplementing the President's FY 2001 budget, pursuant to the Global Change Research Act of 1990. The report describes the U.S. Global Change Research Program (USGCRP); summarizes recent highlights in global change research, assessment, and observations, and highlights of the FY 2001 budget; discusses the seven Program Elements and FY 2001 plans in each of these research areas; and includes an appendix that details the FY 2001 budget, including program components and program highlights for each of the departments and agencies that comprise the USGCRP. Achieving the goals of this program will require continued strong support for the scientific research needed to improve understanding of how human activities are affecting the global environment, and of how natural and human-induced global change is affecting society and ecosystems.

For Further Information:

Environment Division
Office of Science and Technology Policy
Executive Office of the President
Washington, DC 20502
202-456-6202 (voice)
202-456-6025 (fax)
http://www.whitehouse.gov/WH/EOP/OSTP/html/OSTP_Home.html/
http://www.usgcrp.gov/